Critical Acclaim for *Crafting Your Research Future*

"Ling and Yang summarize the practical aspects of the expectations for the modern graduate student. They will all benefit."
—Randy Goebel, Professor of University of Alberta

"It will be tremendously useful to post-docs and graduate students alike (and perhaps even some junior faculty!)"
—Adrian M. Owen, Professor and Canada Excellence Research Chair, Western University

"Want to have a successful research career? This is a nice guide and companion!"
—Jiawei Han, Abel Bliss Professor, University of Illinois at Urbana-Champaign

Copyright © 2012 by Morgan & Claypool

Crafting Your Research Future
A Guide to Successful Master's and Ph.D. Degrees in Science & Engineering
Charles X. Ling and Qiang Yang
www.morganclaypool.com

ISBN: 9781608458103   paperback

ISBN: 9781608458110   ebook

DOI: 10.2200/S00412ED1V01Y201203ENG018

A Publication in the

SYNTHESIS LECTURES ON ENGINEERING

Lecture #18

Series Editor: Steven F. Barrett

Series ISSN
ISSN 1939-5221   print
ISSN 1939-523X   electronic

Crafting Your Research Future

A Guide to Successful Master's and Ph.D. Degrees in Science & Engineering

Charles X. Ling
University of Western Ontario, Ontario, Canada

Qiang Yang
Hong Kong University of Science and Technology, Hong Kong, China

SYNTHESIS LECTURES ON ENGINEERING #18

MORGAN & CLAYPOOL PUBLISHERS

ABSTRACT

What is it like to be a researcher or a scientist? For young people, including graduate students and junior faculty members in universities, how can they identify good ideas for research? How do they conduct solid research to verify and realize their new ideas? How can they formulate their ideas and research results into high-quality articles, and publish them in highly competitive journals and conferences? What are effective ways to supervise graduate students so that they can establish themselves quickly in their research careers? In this book, Ling and Yang answer these questions in a step-by-step manner with specific and concrete examples from their first-hand research experience.

KEYWORDS:

Research Methodology, Research Career Guidebook, Guidebook for Graduate Students, Ph.D. and Masters Research, Writing and Publishing Papers, Writing Ph.D. Thesis, Science and Engineering Career.

Contents

Acknowledgments..ix

Preface xi

1. **Basics of Research** 1
   1.1 What Is Research?.. 1
   1.2 Who Are Researchers and Who Are Not? 2
   1.3 What Do Researchers Do?.. 3
   1.4 What Skills and Abilities Are the Most Important
       for Researchers? 7
   1.5 What Are Some Pros and Cons of Being Researchers?.......... 10
   1.6 So, How Do I Become a Researcher? 12
   1.7 What Are the Differences between a Master's
       and Ph.D. Thesis? 12
   1.8 How to Find a Suitable Supervisor for Ph.D. Study 13
   1.9 How Long Does It Take to Get My Ph.D. Degree? 16
   1.10 Three Representative Graduate Students........................ 17

2. **Goals of Ph.D. Research** 19
   2.1 Goal #1: Be the Best in the Field................................ 19
   2.2 Goal #2: Be Independent 25
   2.3 Three Key Tasks for Getting a Ph.D. (and Master's) Degree 27
   2.4 The Milestones of Getting a Ph.D. Degree..................... 28
   2.5 Distractions to the Goals.. 30

3. **Getting Started: Finding New Ideas and Organizing Your Plans** 33
   3.1 Your First Year... 33
   3.2 How to Find Relevant Papers (Literature Search)............. 35
   3.3 How to Read Papers.. 37

3.4   Where to Get New Ideas 41

3.5   From Ideas to Research and Thesis Topic............................ 44

3.6   How Do I Know If I Am on the Right Track? 46

3.7   Setting Up a Plan for Ph.D. Thesis Early............................ 47

3.8   Learning to Organize Papers and Ideas Well....................... 49

4.   Conducting Solid Research **51**

4.1   An Overview of a Research Process 51

4.2   Jim Gray's Criteria 54

4.3   The Research Matrix Method.. 59

4.4   Carrying Out Your Research 63

4.5   Brand Yourself.. 67

4.6   Empirical vs. Theoretical Research 69

4.7   Team Work and Multi-Disciplinary Research 74

5.   Writing and Publishing Papers **77**

5.1   "Publish or Perish"... 78

5.2   Why Publishing Top-Quality Papers Is Hard 80

5.3   What Makes a Great Paper? 81

5.4   A Few Untold Truths about Research Papers 83

5.5   The Roles of You, Your Supervisor, and Proofreader 85

5.6   Supervisors: How to Improve Your Students' Writing Most
Effectively... 87

5.7   Where to Submit: Conference or Journals?........................ 89

5.8   How Are Full-Length Conference Papers Reviewed? 91

5.9   How Are Journal Papers Reviewed?..................................... 94

6.   Misconceptions and Tips for Paper Writing **99**

6.1   "It's So Obvious that Our Paper Is Great"............................ 99

6.2   "It Is Your Responsibility to Understand My Paper" 102

6.3   The 10/30 Test 103

6.4   Top-Down Refinement of Papers 104

6.5   Create a Hierarchy of Subsections and Choose
Section Titles Carefully... 109

6.6   Tips for Paper Writing.. 110

6.6.1   Use Certain Words to Signal Readers........................... 111

6.6.2   Use Simple Sentences, But Not Simpler....................... 111

6.6.3 Use a Small Set of Terms Throughout the Paper............. 112

6.6.4 Use Examples Early and Use Them Throughout the Paper... 113

6.6.5 Use Figures, Diagrams, Charts, Photos, Tables................ 113

6.6.6 Write about Your Motivations and Justifications 114

6.6.7 Pose Potential Questions and Answer Them Yourself.. 114

6.6.8 Emphasize and Reiterate Key Points in Your Paper 115

6.6.9 Make Connections Throughout the Paper...................... 116

6.6.10 Format Papers for Easy Reading 116

6.7 Other Misconceptions and Flaws.................................... 117

6.8 Summary 118

7. **Writing and Defending a Ph.D. Thesis** **121**

7.1 Thesis and Dissertation.. 121

7.2 Thesis Organization: Top Down or Bottom Up 122

7.3 Defending Your Thesis... 126

8. **Life After Ph.D.** **131**

8.1 A Day In the Life of a Typical Professor 131

8.2 Applying for Research Grants... 135

8.3 Technology Transfer.. 144

Summary..**151**

References **153**

Author Biographies..**155**

Acknowledgments

First of all, we thank the thousands of graduate students and researchers who attended our seminars on how to conduct solid research and how to write and publish top-quality papers. To these students: your enthusiasm for research, your questions, and your suggestions helped us tremendously in preparing and finishing this book. It is our wish to motivate and support more young researchers in their path toward a successful research career.

From Charles Ling: I wish to dedicate this book to my parents Zongyun Ling and Ruqian Cao. They were both life-long, awarding-winning, high-school teachers. They gave me so much inspiration and guidance throughout my youth, which laid the foundation of my research life from my graduate study to a full research career. This book summarizes over 20 years of experience as a researcher, which represents a direct outcome of my parents' love, care, and life-long education for me. I wrote parts of this book on the bedside of my father when he stayed in the hospital for several months in 2011. The book is for you!

From Qiang Yang: I dedicate this book to my parents, Professors Haishou Yang and Xiuying Li, who were professors in China's Peking University and Tsinghua University, respectively, where they worked all their lives until retirement. My parents inspired me to pursue a research career in Astrophysics and then Computer Science. Through their own exemplary research careers, they showed me the "garden of research," which is "full of beautiful flowers waiting to be picked" (to quote from my father). To them: my deepest love and appreciation!

Many colleagues, friends, and students read the early drafts of the book, and made many detailed suggestions and comments. Brian Srivastava, a very capable

Ph.D. student at Western, provided detailed and thorough suggestions and comments in early draft of the book.   Lili Zhao, a research assistant in Hong Kong University of Science and Technology (HKUST), made detailed edits. Luiz Fernando Capretz, Jimmy Huang, Huan Liu, Aijun An, Steve Barrett, Wei Fan, Jenna Butler, Randy Goebel, Eileen Ni, and many others gave us suggestions on various parts of the book.   Adrian Owen, Stephen Sims, C.B. Dean and Jiawei Han gave us encouragement after reading parts of the book. Many of our former Ph.D. students, especially Victor Sheng and Weizhu Chen, gave us comments and suggestions. To all: our special gratitude!

Preface

When we were children, we were often asked, "What do you want to be when you grow up?" Our answer, and dream, was: "I want to become a scientist!" Indeed, we have many giants to look up to: from Newton to Einstein, to Nobel Laureates (such as Richard Feynman). However, when we really went through the process of becoming scientists, we found that the road was not all that straightforward; it was full of misconceptions that took many instances of trial and error to find out. True, there are numerous online articles and commentaries, telling young people how to become a successful researcher and scientist, but most are scattered and some are biased. There has not been a central place where one can find advice targeted to answering such questions about the first and most important stepping stone toward becoming a researcher: graduate study.

This situation will hopefully be changed with the publishing of this book, "Crafting Your Research Future." Summarizing more than 20 years of experience in research and practice, we have put together an accessible introductory book for aspiring researchers in science and technology.

We strive to make this book accessible to a general audience such that it could be used by readers interested in general areas of science and technology. At the same time, we try to be as specific and concrete as possible, to show you step by step, how research is conducted. We also include several case studies and many specific examples.

The book also summarizes over a decade of seminars and talks at over 50 universities and institutes across the world on the topics: how to do research and how to write and publish research papers. In all our talks and seminars, the attendance has been overwhelming, with hundreds of students and faculty filling the lecture hall. Questions and answer periods stretched beyond the talk. Each seminar gives us a chance to improve our presentation, and this book is the latest in this effort.

Here is a brief overview of the book.

Chapter 1 sets the general stage for the book, by discussing what research is, and what researchers are.

Chapter 2 discusses the goals of research and lays down the basic steps toward a research career.

Chapter 3 answers the question of how to get started by looking for a suitable supervisor, reading relevant literature and getting new ideas. It advocates a principled method with three key steps to guide any research effort.

Chapter 4 presents a general methodology on how to evaluate potential directions in order to find one that is high impact, and how to conduct solid and thorough research. It goes into details on the differences between empirical, theoretical, and multidisciplinary research.

Chapter 5 discusses how to write and publish high quality papers in competitive journals and conferences. It also provides perspectives from reviewers of journal and conference articles, and provides tips on how to handle reviewers' comments.

Chapter 6 presents a collection of commonly met misconceptions and pitfalls, and offers useful tips on paper writing.

Chapter 7 discusses how to plan, organize and write a Ph.D. thesis, and how to defend a Ph.D. thesis. It presents two typical approaches: top down and bottom up.

Chapter 8 gives readers a detailed picture on life after Ph.D. by depicting a realistic picture of the life of a professor. This chapter also discusses important issues such as technology transfer.

Both authors have been university professors for over 20 years, and have supervised many Master's and Ph.D. students. Some of our Ph.D. students are now professors in universities in the US, Canada, China, and other countries. Some are researchers in research institutes and research labs at Google, IBM, and other companies. Some are running start-up companies successfully.

We have put a lot of effort into helping our students to choose thesis topics, do solid research, and to publish top-quality papers. Often co-authoring papers with students in our research, we have published over 300 peer-reviewed journal and conference papers. We have also been Associate Editors and Editor in Chief of several top-ranked journals in the computer science field, such as IEEE[1] Transactions on Knowledge and Data Engineering and ACM[2] Transactions on Intelligent Systems and Technology.

A more detailed biography of the authors can be found at the end of the book.

[1] IEEE: Institute of Electrical and Electronics Engineers
[2] ACM: The Association for Computing Machinery

CHAPTER 1

Basics of Research

In this chapter, we examine some of the frequently asked questions about the basics of research. If you want to know what researchers and scientists are like, if you are still not sure whether doing research is a suitable career for you, or if you are just a beginner in research, read on.

1.1 WHAT IS RESEARCH?

In natural science and engineering, "research" can be loosely defined as making *novel* and *significant* contributions to our understanding of the world, based on reproducible observations and verifiable results. There are often large overlaps in scope between research in science and engineering. In science, the emphasis is on discovering new theories, paradigms, approaches, algorithms, simulations, experiments, and so on, while in engineering, the emphasis is to solve real-world problems with new technologies, designs, processes, methods, models, testing, and so on. Novelty and significance are the key ingredients of research in both science and engineering.

Something that is "novel," "new," "original," "creative," "unconventional," "innovative," or "ground-breaking," means that it was unknown previously. In the research arena, it usually means that it was not published previously in academic journals, conference proceedings, books, technical reports, or other accessible venues. Novel work must also be significant or have high impact, be important, useful, seminal, and so on, to qualify as research. A highly significant research will ultimately bring huge benefit to many people in the world.

How can novelty and significance be judged? Clearly, if a work, either in the same or in different forms, has been published previously, it is certainly not novel.

If the work is only a small derivation or obvious improvement from previous work, it is also not considered novel. "Significance" is harder to judge. A research work is highly significant if it lays the foundation for future work, very significant if it discovers new paradigms or methods that can solve new problems, or quite significant if it outperforms previous methods on some important tasks. Often significance can only be assessed in the future; in many conferences and journals, a 10-year "most significant paper award" is given to a paper that made the most significant contributions 10 years ago. One informal measure of the impact of research work is how many other researchers have used this work as the basis to build their own research programs, and how many research papers have cited the work after it was published. Thus, experienced researchers are often called on to review and judge the significance of research papers submitted for publication.

Chapter 3 to Chapter 6 will discuss the process of doing research in greater details.

1.2   WHO ARE RESEARCHERS AND WHO ARE NOT?

Researchers conduct research as the main part of their professional career. They are also often referred to as scholars or academics. This book is written mainly for researchers-to-be, especially Ph.D. candidates, in natural science and engineering. The book may also be helpful to young researchers in their early careers, such as assistant and associate professors in universities. In many of the sciences, especially natural sciences, researchers are also called scientists. In mathematics, they are known as mathematicians; in computer science, computer scientists; in biology, biologists, and so on. Most researchers work at universities, research institutes, and research labs of large companies.

One misconception about researchers and scientists that we want to correct is that typical researchers are old "nerds" or "geeks": people with either gray hair or no hair, wearing torn clothes, thinking all the time and talking to themselves, and who are incomprehensible and unapproachable. In reality, most researchers and scientists are full of life, fun, versatile, and have many hobbies. A typical example is Professor Richard Feynman, a physicist and Nobel Laureate who also played music, learned to draw, went to bars, married (three times), acted in a film, traveled around the world, wrote many popular science books, and so on. We highly recommend that you read his legendary book: *Surely You're Joking, Mr. Feynman!* (published by W.W. Norton). In fact, most researchers and scientists that we know of, including us, are full of color and are "normal" people!

Here are some typical examples of non-researchers. Salespersons are not researchers. People who follow routines and instructions, like factory workers and accountants, are not researchers. Teachers and instructors whose main job is teaching in schools, colleges or universities are usually not researchers, unless they develop, test, and experiment on new techniques for education. Engineers solve real-world problems and build things, and if they make novel and significant contributions in the process, they are also researchers. Medical doctors and dentists are professionals and usually not researchers, unless they conduct research on new medicines or new treatment.

On the one hand, researchers, such as professors, often also do other routine work such as teaching, managing research grants, and so on. On the other hand, many non-researchers also do some research in their work.

1.3   WHAT DO RESEARCHERS DO?

Here is a list of 11 tasks that researchers often do in their day-to-day work. Chapter 8 describes a typical day in the life of a researcher in going about some of these tasks.

Task 1.   Explore and come up with new ideas (i.e., novelty of research). To make sure that the idea is new, researchers must know well the frontier of their research areas by conducting an extensive literature search. Chapter 3 of this book will discuss this in great detail.

Task 2.   Validate and implement the new ideas to see if they work, and make them work well (i.e., significance of research). Chapter 4 will describe how to do solid research in detail.

Task 3.   Write up the research outcome in Tasks 1 and 2 as "manuscripts" or "papers," and submit them to academic journals, conferences, book publishers, or other media outlets. Usually the research manuscript is first peer-reviewed by other researchers in the same field, as they are probably the only qualified people to determine if the work is novel and significant enough to be accepted for publication. To ensure fairness and frankness, reviewers are always anonymous. Often the selection is quite competitive: many top journals and conferences may have an acceptance rate of only 10-20% for full-length papers. That is, the majority of the submissions are rejected. Chapter 5 and Chapter 6 will discuss how to write manuscripts well. Chapter 5 will also share some insights about how papers are reviewed.

Task 4.   Review papers and manuscripts from other researchers to see if they are novel and significant enough to be accepted for publication. In this role, the researchers become reviewers, as mentioned in Task 3.

Task 5.   Administrate academic journals and organize academic conferences, especially when one becomes an established researcher. These duties are often time-consuming, and are usually on a voluntary basis (i.e., no extra income), although in the long run, they help increase the academic and professional visibility of the researcher in his/her field.

Task 6.   Attend academic conferences to present papers, to learn the most recent research results, and to exchange research ideas and knowledge with other researchers. This is also an important "networking" time with other researchers in the same field. Established researchers may also be invited to give keynote speeches at conferences.

Task 7.   Apply for research grants from government and other organizations to support research, including providing support to graduate students. As most research work may not have immediate monetary return, usually the government and organizations invest money as grants to support research. The grant application can also be highly competitive. See Chapter 8.

Task 8. Supervise graduate students, especially Ph.D. candidates, so they become researchers in the near future. This book is mainly about how to successfully complete both Master's and Ph.D. study, including thesis writing and defense. It is also about how to be an effective supervisor.

Task 9. Teach graduate and undergraduate courses in universities and colleges. Some of those students will become researchers in the future.

Task 10. Perform administrative work for the organizations that researchers work in. In the university this includes committee work in the department, faculty, or college, writing reference letters, and so on.

Task 11. Commercialize research outcome (also known as Conducting "Technology Transfer"). If particular research work is fundamental, the research outcome is often published in an open forum and the research outcome is published in an open media, often free for other researchers to verify and utilize for further research. However, some research is applicable in the real world, and the results can be commercialized. In this case, researchers may also seek to protect their intellectual properties with patents and so on. A small number of researchers even create spin-off start-up companies to commercialize their research. We also have a wealth of experience in industrial applications and technology transfer, and we will discuss this topic in Chapter 8 of the book.

Many researchers work as professors, with ranks known as assistant, associate, or full professors, in research-oriented universities. Usually they spend a considerable amount of time and effort in Tasks 1, 2, 3, 7, 8 and 9. In terms of research topics and methodologies (Tasks 1 and 2), university professors are quite free in choosing whatever interests them the most, provided that they can produce novel and significant results. They can also freely express their academic views and results in publications (Task 3), provided that the work is peer-reviewed and accepted. Depending on individual researchers, they can also be quite involved in Tasks 5 and 11, and other tasks on the list.

Researchers may also work for large companies and various organizations. For example, Microsoft, Google, IBM, GE, DuPont, Pfizer, Boeing, National Institutes of Health (NIH), National Research Council of Canada, and so on all have divisions for research. Researcher's main duties usually include Tasks 1, 2, 3, and 11. They may also be involved in Tasks 4, 5, and 8; in fact many Ph.D. students conduct research for companies and organizations as interns. In terms of research topics, they are not as free as university professors; they often have to work on research problems that are directly related to the company's service and products. However, Google has a policy, called Innovation Time Off, where Google engineers can spend 20% of their work time on new projects that the engineers and scientists are interested in doing, but may not be on the company's current agenda.

Ph.D. candidates are researchers-to-be, supervised by university professors. Their main job is Tasks 1–3. Often they also work with their supervisors on many other tasks including Task 8. As senior Ph.Ds, they may help junior Ph.Ds and Master's students to do research. They need to learn to do these tasks so that when they finish their Ph.D. program, they become independent researchers who can perform Tasks 1–11 reasonably well by themselves. Some disciplines, such as Biology and Physics, have the practice in which one often follows a Ph.D. with a Post Doctoral fellowship, which is a half step between Ph.D. students and professors. These positions are usually short, but give the researcher an opportunity to learn parts of the process they didn't see as students, notably managing group budgets and different approaches to management, from their Ph.D. supervisors.

1.4   WHAT SKILLS AND ABILITIES ARE THE MOST IMPORTANT FOR RESEARCHERS?

Both of us (the authors) were fascinated by the stories of the famous mathematicians and physicists (such as Archimedes, Newton, Euler, Gauss, and Einstein), and dreamed of being scientists when we were young. We can still vividly recall the time when we were middle school students in China. This was the time when the Culture Revolution (during which intellectuals were suppressed[1]) had just ended, and economic reform was just about to start. One day, a vivid news report entitled "The Goldbach's Conjecture" (in Chinese), appeared in all major newspapers, radios, and magazines in China. It reported a Chinese mathematician named Chen,[2] who proved the "Last Hurdle" to solving the Goldbach's Conjecture. The report was like "a thunder," awakening the youth in China as the "spring time for modern science" had arrived in China!

The Conjecture, proposed by Goldbach in 1742, can be simply stated as: every even integer greater than 2 can be expressed as the sum of two prime numbers (one prime plus another prime, or 1+1). Much progress had been made toward the proof of the conjecture since it was proposed. For example, mathematicians before Chen proved that any sufficiently large even integer could be expressed as the sum of two primes, or a prime and the product of 6 primes (1+6). Chen spent more than 10 years in his 50 square-foot apartment and proved the "Last Hurdle" of the Conjecture: every sufficiently large even integer can be expressed as the sum of a prime and the product of two primes (1+2). His proof took over 200 pages. There was only one step to reap the "crown in mathematics" (proving 1+1).

One of us (Ling) was so inspired by the report, and intrigued by the apparently simple conjecture, that he went to buy several books on number theory by himself (in those years food was still a scarce resource), and spent days passionately reading them and trying to prove 1+1. He knew the proof must be hard, but he believed that he could be very creative, and could think of a novel approach that

---

[1] http://en.wikipedia.org/wiki/Cultural_Revolution
[2] Chen Jingrun. http://en.wikipedia.org/wiki/Chen_Jingrun

others had not thought of. One day he believed he found an ingenious proof on only two pages, and he rushed to his father announcing his great discovery! His father calmed him down, and asked him to check the proof carefully. You can imagine the outcome.

After he entered his undergraduate study in Computer Science, he wrote a computer program to verify the Conjecture for integers with many digits, hoping to find a counter example to refute the Conjecture. You can imagine the outcome.

The Conjecture of 1+1 remains open today. It has now been verified by computers to even integers of 18 digits long, according to the Wikipedia page on the Conjecture. But he still dreams that someday he can prove it, or refute it.

Do you have dreams, passion, curiosity, and creativity? Almost everyone does. These are some of the most important qualities for being researchers. Besides these, we list what you need to succeed on the path of research below.

- **Passion, focus, enthusiasm, and interests in research.** Researchers must be keenly interested and highly passionate about their research, so that they can stay focused and study research problems deeply over weeks, months, or even years.
- **Curiosity and creativity.** Researchers must be highly curious and creative. They often ask many questions, both in research and in daily life, and think "out of the box" for new ideas and solutions that most people have not thought about. They may often think of "plot holes" when they watch movies, and discuss how to make the movies better!
- **Critical and independent thinking.** Researchers must not just follow the conventional wisdom and established views. They must think independently and ask critical questions (e.g., what is wrong? How can I make it better?).
- **Risk taking.** Often new ideas do not work, or have been explored and published by others. Researchers must be willing to take risks in exploration.

- **High scientific integrity.** Researchers must be willing to take risk and fail, and be honest and transparent about failure. They must also be truthful in reporting their research outcome (see Section 5.4). Peter Medawar's excellent book, Advice to a Young Scientist, should be read by all graduate students.

- **Quick learning and strong analytical and problem-solving skills.** Researchers need to be quick learners for new knowledge and techniques needed for research. They must have good analytical and problem-solving skills (which can be learned) to analyze, implement, and test their new ideas (conducting solid research).

- **Diligence.** Researchers must stay focused, think deeply, and study new research problems in a great depth. They often work for more than 8 hours a day, and 40 hours a week. They may spend months or even years trying to make some breakthrough in their research.

- **Good communication skills.** As researchers disseminate their results via manuscripts or papers, it is extremely important that they learn to write well. They must be a good "story teller": knowing their audience (readers) and convincing others quickly to believe in their story (accepting their papers). They must also be able to present their ideas clearly and convincingly in conferences. University professors usually need to teach courses so they must learn to be good instructors, and love teaching classes.

What we list above are crucial traits and skills for researchers; other positive personal traits and abilities are also important. You can ask yourself if you are suitable to be a researcher by comparing your personality and strengths with the list above. Note that, however, most of the required abilities and skills can be learned and improved, and one weak ability may be compensated by other stronger ones. For example, if you are not very strong in mathematical theory, you can choose a thesis topic on applied or experimental research.

1.5   WHAT ARE SOME PROS AND CONS OF BEING RESEARCHERS?

Here is a list of "Pros" of being researchers. It reflects our personal views.

- **High career satisfaction.** Researchers have more freedom in choosing what interests them the most to work on than many other forms of employment. University professors also have flexible working hours in their university office, and have summer months in which they can focus on research, and travel for conferences and research visits. It is also highly rewarding when researchers make new discoveries and breakthroughs, which benefit people and are recognized by their peers.

- **Protected research environment.** Researchers, especially university professors, have a relatively stable and protected position for research. These positions are usually not much affected by the short-term economic downturns and by the opinions of politicians. To protect academic freedom and give researchers time to produce high-impact results, most universities have a "tenure" system—tenured professors cannot be fired easily when they dissent from prevailing opinion, openly disagree with authorities of any sort, or spend time on unfashionable topics' (Wikipedia), or if their research 'quantity' fluctuates due long-term research. Publishing a few high-impact papers is often better than publishing more low-impact ones.

- **Reasonably good income.** Researchers usually have a Ph.D. degree, and their entry salary is often much higher than those with Master's or Bachelor's degrees. In addition, researchers' salary is also quite stable; there is usually no "bonus" or "penalty" if they accomplish a lot or a bit less for Tasks 1–10 in a particular year. This can be regarded as a disincentive, we suppose, as salespersons can earn more money if they sell more products. However, professors can earn an extra income if they hold secondary consulting or business positions, or if they are successful in technology transfer (Task 11).

- **Researchers are highly respected by most societies.** In 1999 an A&E Biography program selected the 100 most influential persons of the last millennium (from the year 1000 to 2000). Among the top ten most influential persons, five are scientists; they are Galileo (#10), Copernicus (#9), Einstein (#8), Darwin (#4), and Newton (#2). They are highly respected by people. They are also the ultimate role models for young researchers. In addition, we have many other giants to look up to: from Nobel Laureates (such as Richard Feynman), to winners of the Turing Award (the highest distinction in the computer field) and John Fritz Medal (referred to as the highest award in the engineering).

Some possible "Cons" of being researchers are:

- **Researchers often work more than 8 hours a day and 40 hours a week to stay competitive.** The amount of work needed for doing Tasks 1 to 11 well can be daunting and intimidating. However, most researchers are highly effective, and they learn to manage time well. In addition, research is a career, not just a "job." Often researchers are intrigued by research questions, and enjoy thinking about them when they have extra time. They are also thrilled when a breakthrough is made.

- **Long preparation time.** To become a researcher, one usually needs to study at a university for one's Bachelor's, Master's, and Ph.D. degrees. Sometimes they need to take a few years of post-doctoral positions after their Ph.D., before a more stable position is secured. However, Ph.D. candidates and post-doctoral fellows are partially researchers. They usually receive a small to moderate income during that time.

- **Relatively narrow career choice.** Depending on the economy and available funding, in some years the number of job openings for post-doctoral fellows and professors in universities and research-ers in companies can be limited. In addition, the number of Ph.D. students graduating in the last few decades is increasing. Thus, the competition for new positions for professors and researchers can be very high. Also, after you get a Ph.D. degree, you may be "overly qualified" for jobs for Master's (or Bachelor's) degrees. However, the economy goes in cycles, and there are many other career choices for researchers, including entrepreneurship (start-ups). When you choose research topics, you must also think carefully what type of jobs (e.g., professors, researchers in companies, etc.) you want to take in the future. We will discuss how to select research topics in Chapter 3.

1.6   SO, HOW DO I BECOME A RESEARCHER?

Most researchers learn to do research during their Ph.D. study in a reputable university, supervised by a university professor. To be accepted in a Ph.D. program, you normally need to have Master's and Bachelor's degrees in the same or similar areas. If you study for your Master's and Ph.D. degrees under the same supervisor, usually the combined duration would be shorter than when you work under different supervisors.

1.7   WHAT ARE THE DIFFERENCES BETWEEN A MASTER'S AND Ph.D. THESIS?

First of all, many universities do not require all Master's students to write a Master's thesis; often you can get the same Master's degree by the "course option"—taking a large number of graduate-level courses. But if you want to get a taste of doing research, or if you plan to pursue Ph.D. after your Master's study, then taking the "thesis option" is advised.

As for the differences between Master's and Ph.D. theses, it is difficult to draw the line precisely. It certainly does not depend on the length of the thesis! Some folk legend says that the shortest Ph.D. thesis is 14 pages long, but the longest is over 1,000 pages. Both Master's and Ph.D. theses require novel and significant contributions, but such requirements for a Ph.D. thesis are much higher. Occasionally there is a debate and discussion among thesis examiners if a thesis being examined should really belong to the other category!

One intuitive and approximate way to distinguish between these two types of theses is that the work in a Ph.D. thesis should be publishable or already published in several top journals and conferences in the field. A Master's thesis, on the other hand, may only be publishable or already published in 1-2 medium competitive journals or conferences in the field.

Several typical differences between Master's and Ph.D. theses and students are outlined below:

- A Master's thesis can apply previous work to a new problem or application; a Ph.D. thesis normally contains significantly new theory, methods, and applications.

- A Master's thesis can make small, incremental improvements from previous work; a Ph.D. thesis usually presents and studies a new topic in the field, and makes much larger contributions.

- A Master's thesis can be a critical survey of existing works, often with additional comparison of these works theoretically or empirically; a Ph.D. thesis must present some new methods, and compare them with previous works convincingly either by theory or experiments.

- A Master's thesis can report negative results of a seemingly promising approach, and analyze why it does not work; a Ph.D. thesis should contain positive results in addition to analysis of the negative results.

- If one graduates with a Master's degree, he or she may not yet learn to become an independent researcher; his or her job is usually not doing full-time research. A Ph.D. graduate, on the other hand, should be an independent researcher, and can compete for and take a university professor job directly.

Again, the distinction is vague, and varies among different fields and disciplines, universities, and countries.

Though this book is written mainly for Ph.D. candidates in Science and Engineering, it applies to Master's candidates as well with smaller requirements.

We will indicate special aspects for Master's candidates in appropriate places in this book.

1.8   HOW TO FIND A SUITABLE SUPERVISOR FOR Ph.D. STUDY

Different departments and universities have different ways of pairing graduate students with supervisors. Choosing an appropriate supervisor for Ph.D. study is crucial, as you may be "stuck" with him or her for the next 3–5 years. This is also important but less critical for Master's students as they usually work with their supervisors for 1–2 years. If you and your supervisor do not have a good working relationship in these years, disagreement, tension, and occasionally, hardship and conflict will arise.

In general, the following factors should be taken into consideration when choosing your supervisors:

- **Research areas and interests.** The number one most important factor is whether or not your research passion and interests match well with your supervisor or not. Nowadays professors' profiles, research areas, and publications are usually posted online. We recommend that you go through the department's website of the school that you are applying to carefully, and spend time studying potential supervisors' research and publications. You can also write to the potential supervisors, with some intriguing questions about their research, and ask them if they are willing to supervise you. You should mention your research interests and professors you wish to be your supervisors in the application materials. If a professor agrees to supervise you, the chance of being accepted is much higher, given that you satisfy the basic requirements for the university and department that you are applying. Some departments let students choose supervisors one year after they are admitted. This would give you ample time to take courses offered by various professors and talk to them before you choose a supervisor.

- **Supervisory philosophy and style.** This is almost as important as research areas and interests mentioned above. Two orthogonal dimensions can be considered: the amount of pressure and guidance. Some professors are very demanding, which may not be bad for you if you need some pressure to excel, while others are relaxed, so you work at your pace. On the other hand, some professors give their graduate students a lot of guidance, such as what research problems to work on, how to write papers, while others give them a lot of freedom; you may need to figure out a lot of things, including thesis topics, by yourself. Does a potential supervisor's style of supervision match your personality and working habit? It is important to find this out before settling down with a supervisor. Bring Figure 1.1 to your potential supervisor, and see where he or she stands.

- **Funding availability.** This is also a practical issue to consider. Though the stipend for Ph.D. candidates is usually quite small to live lavishly, you need to be able to live healthily without the need to work eight hours a day in restaurants to make ends meet. Some professors' fund-

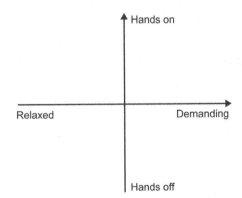

FIGURE 1.1: Two dimensions of the supervisory style: the amount of pressure vs. guidance.

ing can last beyond the usual period, e.g., four years in many universities. Some will last only one or two years. Some professors can fund your conference trip if your paper is accepted, while others may not.

- **Academic and industrial social networks.** It is often said that one is in some ways defined by his or her friends. This is very true for graduate studies and even research careers. Remember that when you join a supervisor's research group, you enter a social network, consisting of all of your supervisor's former students, advisors, current colleagues, contacts and current and future students. These people will be there to help you succeed. This means that you will be able to talk to them easily about new ideas, references, and job possibilities during and after your Ph.D. study. You might leverage on their social networks as well, thus expanding your views, support, and career opportunities.

- **Track record.** It is extremely important for you to do background study to see the academic standing of your potential supervisors in their respective field of research. Today this is easy to do by checking on the publications and their citations in search services such as Google Scholar. It also helps to talk to former and current graduate students of the potential supervisors for the issues discussed above. It is also a good idea to read publications of graduate students under supervision of potential supervisors.

1.9   HOW LONG DOES IT TAKE TO GET MY Ph.D. DEGREE?

We often hear students asking: "How long do we need to take before I can graduate with a Ph.D. degree?"

Well, to us, the student is asking the wrong question. To begin with, we must make clear what a Ph.D. degree entails (for more details, also see the next chapter: "Goals of Ph.D. Research"). Getting a Ph.D. degree means that you are to learn to view the world in a different way, to be able to identify a new and challenging problem on your own, and to solve the problem via a new methodology. It also means that you can claim to the world that you are an expert in one important, albeit narrow field, where others can consider you a leader in this field. One practical criterion is to go to conferences or workshops, where if you find people looking for you because they want to talk to the foremost experts in this field, you know you are ready to graduate, since you have crossed the line to being an expert in your chosen field. Likewise, if you find your work starting to be cited by peers, it is a good sign that you are ready to get your Ph.D. degree, and be called a "Dr"!

Thus, the answer to your question ("how long does it take for me to graduate") is "it depends on you!" Unlike undergraduate study, which usually has a fixed length, Ph.D. study can last from 3 years to 10 years. It all depends on how effectively you accomplish Tasks 1-3 mentioned earlier in this chapter, how you select a suitable Ph.D. topic, and how well you work with your supervisor. This book will provide you with useful guidelines and effective methods of obtaining a Ph.D. in relatively short time. Most of our Ph.D. candidates get their degree in 3 to 5 years.

1.10   THREE REPRESENTATIVE GRADUATE STUDENTS

Throughout this book, we will use three fictitious Ph.D. students as examples. They represent three general types of Ph.D. students, and are based on several past Ph.D. candidates under our supervision.

Student A is a Ph.D. candidate whose strength is academic research and whose career dream is a professorship in a university in the US or Canada. He published a good number of papers in competitive conferences and reputable journals, and got his Ph.D. in 4 years. He then became a post-doctoral fellow for two years to deepen his research. He is now a tenure-track assistant professor in a US university. Through the book we will see how he started, how he chose and tried a few different research topics before settling down on his Ph.D. topic, how he did his research, and how he published many papers.

Student B is a Ph.D. candidate whose strength is empirical research. He wants to work in a research lab in a company. He has written several conference papers and one journal paper related to the topic of his thesis, but more importantly, he has participated in an international competition, and got into the top-3 in a well-known large-scale experimental contest based on real-world data.

Student C is a Ph.D. candidate whose strength is engineering and industrial applications, as well as commercializing his ideas. His research is more application oriented, and his aim is to build some novel and useful systems in the future and start a venture to commercialize his ideas. He developed a novel experimental system to offer sufficient value over the current state of the art and business. Thus, he has successfully filed a patent and published a few news articles on his technology. He also published a few conference papers. He builds a startup business after graduation.

To cover a whole spectrum of graduate students in our book, we refer to these types of students as Student A, Student B, and Student C, respectively, where they correspond to the three different approaches to Ph.D. study: Student A for an academic and fundamental research type, Student B for an empirical-research type, and Student C for an application and entrepreneurial type. Later in the book, we will use Student A1, Student A2, Student B1, and so on, to describe specific instances of various student types to highlight some specific details. In such cases, Students A1 and A2 are two examples of type-A students.

·   ·   ·   ·   ·

CHAPTER 2

Goals of Ph.D. Research

Before we describe in detail how to get a Ph.D., we want to first discuss the ultimate goals of Ph.D. research in this chapter, so that you "begin with the end in mind" for your Ph.D. study.

2.1   GOAL #1: BE THE BEST IN THE FIELD

The number one goal of your Ph.D. study is that, in an important, albeit narrow field (i.e., your thesis topic), *your research is the best in the world* (in terms of novelty and significance) at the time when you complete your thesis, and you are among the very few best experts in that topic.

We can use a very simple figure to illustrate this, as in Figure 2.1. The x-axis represents a certain domain, such as artificial intelligence or water purification. The y-axis is the expertise of knowledge in that domain. A horizontal line in the figure represents the current best knowledge, which is also called the frontier or the state-of-the-art of research. Assume that you will get your Ph.D. in four years. Your yearly research progress is depicted in the figure. Note that this overly simplified diagram is for illustration purpose only.

Near the end of your Ph.D. study (say, the third and fourth years), you should produce research that advances the current best knowledge (so that the "best line" will be raised a bit due to your work). The height of your knowledge curve above the "best line" is roughly the novelty of your work; the width is roughly the significance or impact. You should establish yourself in this important (albeit narrow) field. In other words, you should "own" an area. How do you prove that you are ready to establish yourself in your field? A typical way is publishing a

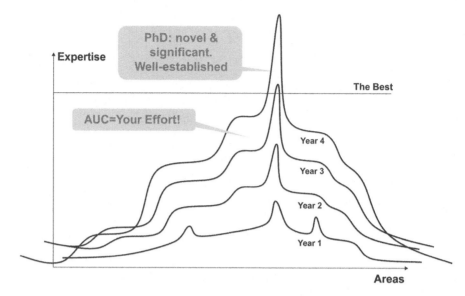

FIGURE 2.1: Progress in four years of Ph.D. study.

series of high quality papers in top-ranked conferences and journals in your Ph.D. area (Chapter 5 and Chapter 6), and successfully write up and defend your Ph.D. thesis (often a nerve-racking process; see Chapter 8).

Student A1, for example, researched in the area of machine learning, an area in Artificial Intelligence (which is an exciting research area that explores what makes us intelligent and how to make machines intelligent as well). He published over 10 top journals and conference papers, and finished his Ph.D. thesis in 4 years. Student B1 published 5 top-ranked conference and journal papers in 4 years, and finished a system that implements the ideas in these papers. Student C1, on the other hand, finished 3 papers in conferences, but also filed one patent successfully. He graduated with a venture-capital company's offer to start up a new business.

Early in your Ph.D. study (e.g., first year), you can explore various topics that interest you (the several small peaks in Figure 2.1). But soon you should focus (one peak), and try to conduct research in one topic in great depth. Examples of a thesis topic can be cost-sensitive learning with data acquisition (Student A1's topic, an area of research in machine learning), active-learning and user interface design for Student B1 (another area of research in computer science), and a new method for harnessing crowdsourcing power (an area of computer and social sciences where computer and human problem solvers work together to tackle a difficult problem, such as language translation) and for Student C1, turn it into a well-formed business model. Chapter 3 and Chapter 4 will discuss this crucial step of selecting Ph.D. topics in more detail.

During the four years of your Ph.D. study, you are putting in efforts to literally "push up" your knowledge curves as shown in Figure 2.1. The Area under Curve (AUC), as shown in the figure, is the area between each curve (such as your knowledge curve for year 4) and the x-axis. The value of the AUC roughly represents your effort. This value is cumulative, and should be steadily increasing during the years of your Ph.D. study. *If you do not put in sufficiently large effort, it is impossible to advance the frontier of the research.* In Chapter 1 we emphasize that researchers must be diligent.

You do not need, and often it is impossible to be highly knowledgeable in all topics in a field. However, you do need to be so in areas related to your Ph.D. topics. The shape of the knowledge curves in Figure 2.1 illustrates this. For example, Student A1's topic is cost-sensitive learning. He is also very knowledgeable on this topic (a form of supervised learning), but less so on other areas of Artificial Intelligence, such as unsupervised learning.

If you are a Master's student in the thesis option, it usually takes two years for you to finish your thesis. The situation is similar to the first two years in Ph.D. study, with two exceptions:

In your first year you often have less chance to explore several topics deeply. You and your supervisor would quickly decide a research topic together, and you would quickly make some novel contributions (your Master's thesis) in a narrow field in the second year. Figure 2.2 (a) depicts this situation.

You may conduct a critical review of several related topics, or write a comprehensive survey of a recent topic. You provide new insights and draw new relations in one or several broad areas. Figure 2.2 (b) illustrates this situation. See

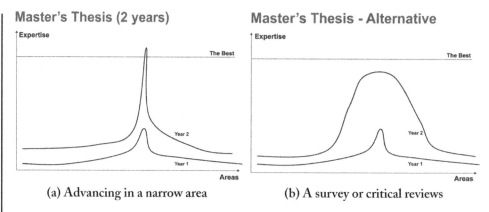

(a) Advancing in a narrow area   (b) A survey or critical reviews

FIGURE 2.2: Progress in two years of Master's study.

Chapter 1 for more discussions on the differences between Master's and Ph.D. theses.

There are several common situations where Ph.D. candidates do not keep this goal in mind, and may have a hard time, or take many more years to finish their Ph.D. theses. We will discuss them briefly below.

In the first situation, they spend most of their Ph.D. study exploring different topics in the field. They may put in a huge amount of effort over four years (high AUC), but they do not have a focus, and do little in advancing the state-of-the-art in any topics they explore. They essentially accomplish several Master's theses. But Master's theses in different topics usually cannot be equal to one Ph.D. thesis. Figure 2.3 depicts this situation.

In the second situation, they explore several different topics, and they manage to make minor contributions and publish a few minor papers in those topics. In this case, they have published on many topics (which may also earn them many Master's theses), but *they do not establish themselves in one*. They do not "own" an area, and other researchers do not remember what they have done. Their Ph.D. theses jump between several different topics, and they may have trouble passing the thesis defense. Figure 2.4 depicts this situation.

Again, your efforts should have a focus. Instead of publishing three papers on three different topics, it is much better to publish three papers on one topic,

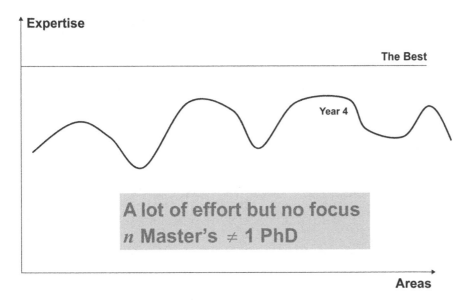

FIGURE 2.3: What to avoid in a Ph.D. study: cover many areas shallowly.

making novel and significant contributions in that area (as in Figure 2.1). By the end of the Ph.D. study, you should "own" an area, so that people will turn to you for advice or for reviewing papers in that area.

A few exceptionally strong and productive Ph.D. candidates do manage to advance research in more than one area, but their Ph.D. thesis usually only consists of published papers in one area. A Ph.D. thesis should be coherent, and focus on one (relatively broad) topic. For example, Student A1 in his early research explored the area of co-training, and published papers in a couple of top confer-ences. But then he found that it was not easy to select a Ph.D. thesis topic. After many brainstorm meetings (see Chapter 3), he changed his area to cost-sensitive learning with data acquisition, a less-studied but important new area of cost-sen-sitive learning. He worked very hard, and published several top-rated papers on this new topic in the next 2.5 years, and established himself well in this area. At the end of his four-year Ph.D. study, he successfully defended his Ph.D. thesis on

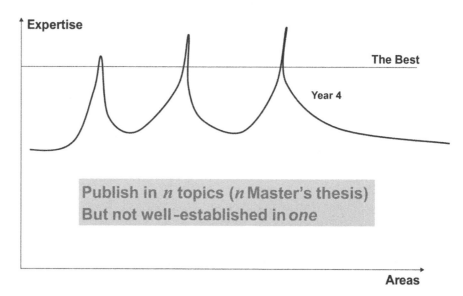

FIGURE 2.4: What to avoid in a Ph.D. study: publishing in several topics but not well established in any one of the topic areas.

cost-sensitive learning with data acquisition. The thesis does not include his early published work on co-training.

Of course, when you become a university professor after your Ph.D., you can develop several strong research areas, working with Ph.D. candidates who have different interests. It is usually quite difficult for a Ph.D. candidate to do so in a short amount of time (3–5 years).

In the third situation, some Ph.D. candidates are keen to find some research topics that can be advanced with little effort (low AUC). They wish they could just focus on a very narrow problem. Figure 2.5 illustrates this situation. This may only be possible if it is an easy and quick extension of your supervisor's (or previous Ph.D. candidate's) work. But in this case, you are not learning to be an independent researcher (see Section 2.2), your contribution is usually minor, and you do not own the area. Thus, for Ph.D. research, the situation in Figure 2.5 is

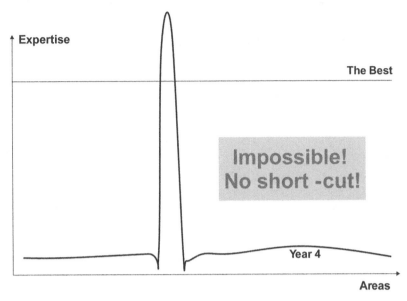

A Good PhD Thesis with Little Effort?

FIGURE 2.5: An impossible situation in Ph.D. study.

impossible. There is no shortcut in producing highly novel and significant work with little effort, or without knowing related areas.

Your supervisor may have a good vision on which research topics and problems are promising to work on as your Ph.D. topic. But your passion, interests, and technical strengths are crucial in Ph.D. topic selection. Chapter 3 and Chapter 4 will explain this in great detail.

2.2   GOAL #2: BE INDEPENDENT

The number two goal of your Ph.D. study is to learn to *become an independent researcher* so that you can do tasks listed in Chapter 1, especially Tasks 1–3 (briefly, find new ideas, do solid research, and publish top papers), well by yourself. This is simply because after you earn your Ph.D. degree, you may be employed in a research-oriented university, and suddenly, you have no research supervisors! You must do everything (Tasks 1–11) by yourself. Being independent also means that

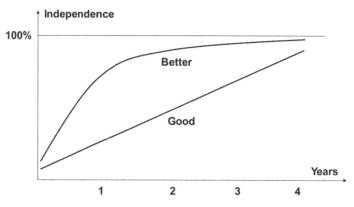

FIGURE 2.6: The "independence curve" for a Ph.D. student.

you become an independent thinker—you have your own academic viewpoints and you can back them up with your research.

We will use another simple figure to illustrate your learning curve of becoming independent. This time the x-axis is the years you spend in the Ph.D. program, and y-axis is the level of independence. It would be good if your independence curve increases linearly with the year, and better if it grows exponentially. See Figure 2.6 for illustration.

To be highly independent, you need to develop quickly those skills and abilities crucial for researchers discussed in Chapter 1.

For all of the three example Ph.D. candidates we described in Chapter 1, they became independent quickly in 4–5 years of their Ph.D. study. More specifically, in the first two years, we (i.e. supervisors) usually held daily or weekly meetings with them discussing new ideas, methods, and experimental results in detail. We also spent a great effort training them to write well their first papers (see Chapter 5). In the later years, we usually met weekly, or whenever needed, to discuss ideas, approaches, and results at high-level. They can extend and modify the ideas, design and modify experimental settings, and write up the paper drafts, mostly by themselves. They have also learned to review papers and write parts of grant proposals. Some of them have also taught courses, and helped supervise Master's students.

2.3 THREE KEY TASKS FOR GETTING A Ph.D. (AND MASTER'S) DEGREE

As we discussed in Chapter 1, as a Ph.D. candidate, your key tasks are Tasks 1–3. We will refer to them briefly as: "find new ideas," "do solid research," and "publish top papers," respectively. It is unlikely that you can get a Ph.D. by doing Task 1, then Task 2, and then Task 3 in a sequence only once. Instead, it is an *iterative* process. There are also often interactions among Tasks 1, 2, and 3. Through several iterations, you broaden and deepen your Ph.D. topic by building up enough research results and publishing enough papers. In this case your Ph.D. thesis would simply be a "recompilation" or integration of the published papers (more about this in Chapters 5–7). Figure 2.7 illustrates this process.

As mentioned earlier, a good rule of thumb is that you have published a good number of papers in reputable journals and conferences based on your Ph.D. topic. In the following four chapters (Chapters 3–6) we will describe strategies for accomplishing these three key tasks effectively. Chapter 7 will provide you with guidelines on how to write your Ph.D. thesis, and successfully defend it.

Note that although your main job as a Ph.D. candidate is Tasks 1–3, often you also work with your supervisor on all other tasks described in Chapter 1. By the end of your Ph.D. study (hopefully within 3–5 years), you should have made

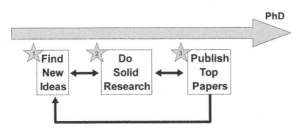

FIGURE 2.7: An iterative process of three key steps in getting your thesis.

significant contributions in your Ph.D. thesis, and established yourself as a young, independent, and competent researcher!

If you are a Master's student, however, you usually only need to go through one, or at most two, iterations of Tasks 1–3, after you take the required courses. You should try to publish 1–2 medium competitive conference papers, which should constitute the main body of your Master's thesis. There is also usually a defense process, but the requirement of novelty and significance in Master's thesis is usually much less than a Ph.D. thesis. Thus, this book is also suitable for Master's students—the basic processes and strategies are the same.

2.4   THE MILESTONES OF GETTING A Ph.D. DEGREE

After discussing the three key tasks in getting a Ph.D. degree, we will describe a series of milestones in 4–5 years of your Ph.D. study. They are:

- Complete all course credits (one to two years).

- Pass a written or oral Ph.D. qualifying exam (~2nd Year).

- Pass a written or oral Ph.D. proposal defense (~3rd–4th year).
- Pass the final Ph.D. thesis defense (~4th–5th year).

A typical graduate school requires a Ph.D. student to complete between five to seven graduate-level courses. For students transferred from other departments and disciplines, they often have to make up some core undergraduate courses that may range from one to three. For example, a Physics student transferring to Computer Science may be required to take Data Structures and Algorithm Design, Theory of Computation, and Operating Systems. All the course works are expected to be finished in roughly one to two years. To prepare yourself for the research, you should try to select the courses that fall within or are closely related to the research area that you are interested in. This is also the time to get your feet wet in the research world. See Section 3.1 for more details.

Many universities require Ph.D. students to take an exam known as the Ph.D. qualifying exam. It can be oral, written, or both. The purpose is to make sure that Ph.D. students have sufficient background knowledge to start and conduct research.

Research life truly starts when you finish the Ph.D. qualifying exam. By this time the student is usually in a celebratory mode, joining parties and sleeping late, mindful that the courses and exam days are finally over, for life! This period in your graduate life may be intoxicating. The problem is that many graduate students may also never wake up from this period. A real challenge for Ph.D. students is that what lies ahead is truly critical, and it is highly advised that the student be prepared for it.

The next milestone is the thesis proposal defense. Ph.D. students need to find a thesis topic (see Chapter 3 and Chapter 4), conduct a comprehensive survey of existing works, find a new and exciting problem, and propose solutions to it. In the thesis proposal, the student is asked to lay out the problem to be tackled, all previous approaches in solving the problem, a detailed discussion of their merits and weaknesses, and the student's particular plan for tackling the problem. The defense is usually oral, in front of a small committee often consisting of faculty members of the same department.

Then you must conduct solid research to prove (or disprove) the proposed solutions, often after many revisions along the way. This task turns out to be more difficult and challenging than most people would think, because it is typically loosely structured, and requires methodical and personal organization and perseverance to handle it well. Again refer to Chapter 3 and Chapter 4 on how to find a good research topic, including Gray's Criteria and the Matrix Method for finding ideas. It is not atypical for a student to try different ideas to see how they pan out, each followed by a conference or journal article. Refer to Chapter 5 and Chapter 6 for paper writing. These activities are basically the iteration of the three key tasks (find new ideas, do solid research, publish top papers) mentioned earlier in this chapter.

After you publish a good number of journal and conference papers on the thesis topic, with your supervisor's approval, you can prepare your Ph.D. thesis and the final defense. Chapter 7 provides a lot of detailed guidelines in writing and defending your thesis successfully.

In addition to accomplishing those milestones, Ph.D. students in most universities also need to spend 10–20 hours per week in a teaching assistant (TA) job.

2.5   DISTRACTIONS TO THE GOALS

Sometimes Ph.D. candidates are distracted during their Ph.D. studies to achieve the two major goals of Ph.D. research effectively.

Many Ph.D. students come from countries whose mother tongue is not English, or whose culture is very different from the country they are enrolled in the graduate study. They may have a language barrier, culture shock, homesickness, and other circumstances (as we did when we first came to the US to study over 25 years ago). Making friends of various cultural backgrounds and quick learning and adaption are important and helpful in minimizing distractions and focusing on your Ph.D. research quickly. If you are not fluent in spoken English (assuming English is used in lectures and paper writing), preparing the lecture materials before lectures is very effective in helping your learn the English language and describe things clearly. Chapter 5 will discuss how to write papers well in English, or any other languages.

In a typical university, a graduate student often works as a Teaching Assistant (TA) to help fund the study. This is like a "real" job in which graduate students are required to perform 10–20 hours in running tutorials, marking assignment and exam papers and holding office hours each week for undergraduate courses. TA work is important and should be accomplished well since besides providing funding for your study, it also gives you an opportunity to learn to present yourself well (to students in the class or to individual students). This can be very useful especially if you want to become a professor in the future. This also presents a challenge because now you have to learn to balance different parts of a graduate-student life: teaching and researching, among others. Students should learn to perform their TA duties effectively, and ensure a sufficient amount of time to be reserved for research. This need to balance one's life is the same if one has to raise a family, learn how to drive, travel for fun, participate in sports and social activities, and generally, enjoy one's life. Indeed, you need to have a life, or as people say, "work hard, play hard."

Even with a high demand in research, Ph.D. students (and faculty members) must find time to do physical exercise to keep themselves fit and healthy. This also improves the efficiency of their brains. In one word, being healthy and staying highly effective in research work and living a good life is the key to achieving the two goals (that is, being the best in your field and being independent) of Ph.D. study successfully in 3–5 years.

Some Ph.D. candidates do work hard, and enjoy doing research, to the point that they do not care how long it takes to get a Ph.D.. They enjoy the Ph.D. program even if it takes more than 8 years. Often the funding from the supervisor and university runs out, and they have to teach many courses to make a living (another distraction). Remember, Ph.D. study is a training process to establish yourself in some area and become a researcher. This can usually be done in 4–5 years, if you follow the guidelines in this book. Then you should defend your Ph.D. thesis, and try to become a full-time researcher, such as university professors, researchers in companies, and be successful.

. . . .

CHAPTER 3

Getting Started: Finding New Ideas and Organizing Your Plans

Let us start from the very beginning of the journey: the first year of your Ph.D. study when you will probably take graduate courses, and maybe a comprehensive exam (or Ph.D. qualification exam). We will focus on issues related to your research, and show you how to start doing research in the first and second years.

3.1 YOUR FIRST YEAR

You should begin your Ph.D. (and Master's) study by taking several broad, graduate-level courses that are related to your future research, especially courses taught by your supervisor. For example, if your research direction is machine learning, then courses such as Machine Learning, Data Mining, and Statistical Learning are highly relevant. If you are willing to pursue a career as an engineer, then you should probably want to take some fundamental engineering courses, including supporting courses such as statistics and probability, computer software tools such as R or MATLAB, and engineering design courses. In addition, you can read by yourself other broad, graduate-level textbooks and take online video courses (there are many good ones) in the field. This will prepare you to have enough basic knowledge for the research in the field. We use a simple figure to illustrate this, as in Figure 3.1. Step A in the figure shows that, after taking courses and reading textbooks, you now have a broad knowledge about the field.

While you are taking graduate courses and reading textbooks, you may find certain topics (such as semi-supervised learning, clustering) fascinating and interesting. Explore them further by:

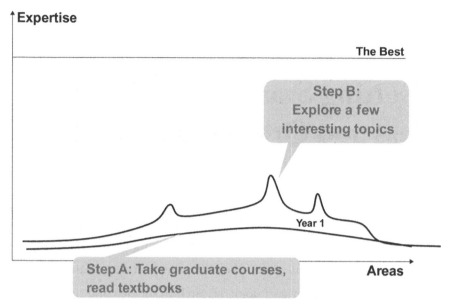

FIGURE 3.1: Progress in the first year: taking courses (Step A), and then exploring a few areas (Step B).

- talking to your supervisors, who may suggest to you certain papers to read, or certain promising directions to explore further;
- trying to find recent survey papers on these topics;
- paying attention to invited speeches, special issues of the journals on these topics;
- trying to go to a conference or two, even low-impact ones that are nearby (and cheap). Talk to researchers, ask them their opinions on the research areas you are interested in, and listen to their paper presentations on what sorts of problems they work on, and how difficult they are. A good paper does not have to win a Nobel Prize to be good.
- finding recent and high-impact research papers (see Section 3.2) and read them critically and creatively (see Section 3.3).
- writing a survey or review paper in the topic area you are interested.

By the end of your first year, and early in the second year, you should have studied a few potential research topics and areas more deeply. Step B in Figure 3.1 illustrates your knowledge curve with a few peaks (the same as the first-year knowledge curve in Figure 2.1).

3.2   HOW TO FIND RELEVANT PAPERS (LITERATURE SEARCH)

With the Internet, Google Scholar,[1] Arnetminer,[2] Microsoft's Academic Search,[3] and other search engines for scholarly publications (such as *Elsevier's Scirus* and *CiteseerX*[4]), and publishers' digital libraries to which many universities subscribe, searching and downloading papers is often a few clicks away (OK, maybe more than a few clicks). The question is: how to find recent and high-impact papers? The process of finding out previous published work is often called literature search. The following are some guidelines:

- Usually, highly reputable journals and conferences review and select papers very rigorously. Thus, papers published there are usually more novel and significant. Your supervisor should know well the reputable journals and conferences in the area, and after you work in the area for a while, you will too.

- One way to judge the reputation of a journal is to look up its impact factor. The impact factor (IF) of a journal is roughly the average number of citations an average paper published in the journal receives in the preceding two years. For example, if a journal has an impact factor of 10, it means that on ave rage, in the past two years, each paper in the journal received roughly 10 citations from other papers published in journals belonging to a reasonably high-quality pool. This pool of journals, known as the Science Citation Index or SCI, is maintained

[1] http://www.scholar.google.com
[2] http://www.arnetminer.org/
[3] http://academic.research.microsoft.com/
[4] http://citeseer.ist.psu.edu/

by an organization called the Institute for Scientific Information (ISI).

- Papers published by senior and well-established researchers often have high impact.

- Use scholarly search engines (mentioned earlier) and search for papers with relevant keywords directly. Those search engines often return a ranked list, combining publication date (novelty), citation counts (impact), and so on together. However, they often do not reveal their ranking algorithms, so you should try different filters, and with several different scholarly search engines.

Figure 3.2 shows an example of searching for the topic of "hierarchical classification" in Google Scholar. It returns a ranked list of papers based on an unknown ranking algorithm, but it seems heavily based on citation counts. We can see that the first few papers were published over 10 years ago, with hundreds

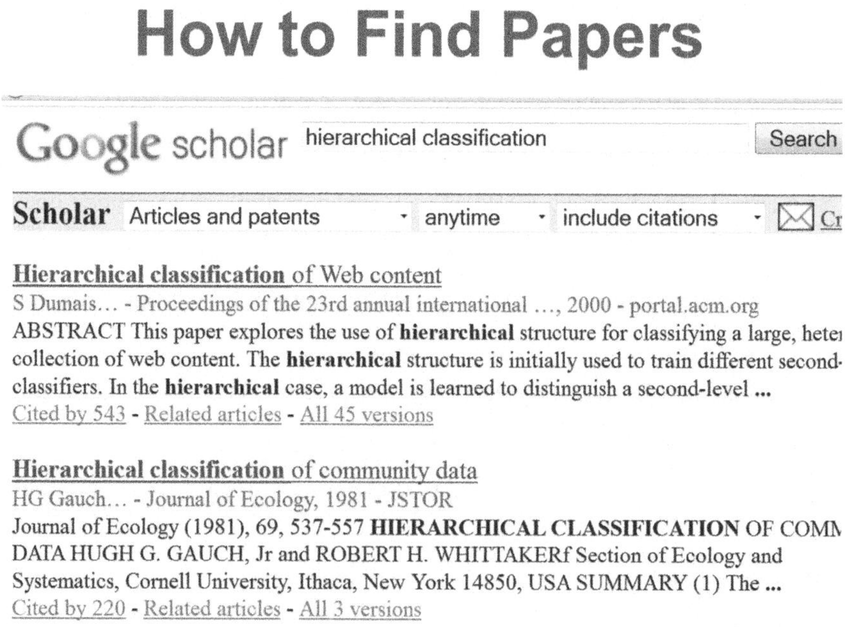

How to Find Papers

Google scholar   hierarchical classification   Search

Scholar   Articles and patents   · anytime · include citations · ✉ Cr

Hierarchical classification of Web content
S Dumais... - Proceedings of the 23rd annual international ..., 2000 - portal.acm.org
ABSTRACT This paper explores the use of **hierarchical** structure for classifying a large, heter
collection of web content. The **hierarchical** structure is initially used to train different second-
classifiers. In the **hierarchical** case, a model is learned to distinguish a second-level ...
Cited by 543 - Related articles - All 45 versions

Hierarchical classification of community data
HG Gauch... - Journal of Ecology, 1981 - JSTOR
Journal of Ecology (1981), 69, 537-557 **HIERARCHICAL CLASSIFICATION** OF COMM
DATA HUGH G. GAUCH, Jr and ROBERT H. WHITTAKERf Section of Ecology and
Systematics, Cornell University, Ithaca, New York 14850, USA SUMMARY (1) The ...
Cited by 220 - Related articles - All 3 versions

FIGURE 3.2: An example of searching for a topic with Google Scholar.

of citations (results in Figure 3. 2 were obtained on June 9, 2011). But how to find recent papers which often have few or no citations? Google Scholar offers several useful filters, especial the time filter, with "anytime" as default. You can change the time to be "since 2010," for example, and it returns a ranked list of papers since 2010.

After you find some recent and high-impact papers, read them. You can then search again with different keywords used in those papers. For example, "text categorization," "taxonomy learning," and "faceted browsing" are different terms for "hierarchical classification." You can also use the "Cited by" feature of Google Scholar, and find all papers that have cited a paper. Pretty soon you will find tens or hundreds of papers on the topic that you are interested in.

How can you read all of these papers, many of which can be quite challenging to read?

3.3   HOW TO READ PAPERS

You might ask: who does not know how to read papers? Just read them like reading textbooks! Indeed, most of us read tons of textbooks. Well, there is a fundamental difference between reading well-explained textbooks and reading research papers. You read textbooks mainly for understanding and learning, expecting all details are worked out and "fed" to you, whereas when you read research papers, you are expected to develop your own research ideas and questions to *surpass* their authors! Another simple way to see the difference is that you read textbooks for *receiving* existing knowledge, while you read research papers for *outputting* new knowledge. Clearly, the objectives of these types of reading are entirely different.

We often see new Ph.D. (and Master's) students take two "bad" approaches in reading research papers. The first bad approach is to spend too much time reading the paper very carefully and trying to understand every detail of it, as they read textbooks. Some new Ph.D. students spend days reading one 10-page conference paper, and come to ask us, "how is this done exactly?" We sometimes receive emails asking us implementation details of our published papers. (However, if other researchers are trying to replicate our published work, we will try our best to reply to them.)

The right approach is to quickly understand the problems, assumptions, main ideas and the proposed solutions presented in the paper at a high-level. You will then spend more time to analyze the paper critically, find new ideas, and conduct research to surpass the previous papers.

More specifically, you should probably spend 30% (or less) of the time reading and understanding the paper's main idea and solutions. This amounts to perhaps 1–2 hours to read a 10-page conference paper. How can you read quickly to understand the main idea of a paper, which can be quite technical? Well, the structure of the paper, if well written, should already give you a hint as to which parts are important to read. A research paper usually consists of an abstract and the body of the paper, which is divided into sections with section headings. Each section may have subsections, and then subsubsections, and so on (we will discuss the paper structure more in Chapter 6). A well-written paper should be structured in a top-down fashion; that is, the high-level ideas and solutions should be described in high-level (sub)sections. Thus, when we read papers, we should focus on high-level (sub)sections, and skim over low-level ones which usually contain technical details.

Figure 3.3 illustrates a hypothetical paper and its associated structure. Quite often you only need to read the Abstract and Introduction sections quickly (say 15 minutes) to determine whether the paper addresses your thesis topic. If not, there is no need to read it further. If yes, you would continue to read the main framework of the proposed algorithms, main results of experimental comparison, and conclusions. We use red boxes to denote the high-level sections you would read more carefully. The parts marked with the yellow boxes may be quickly skimmed. The parts without the boxes may be omitted in the initial reading. This helps us to read a research paper quickly for the main ideas, problems, and solutions at the high level.

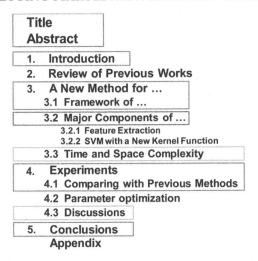

FIGURE 3.3: Structure of the hypothetical paper. We suggest you read the high-level sections (red box) more carefully, skim the lower-level sections (yellow box), and omit technical details (unboxed).

Understanding how a paper should be structured and written will also help you learn to write your papers this way in the future. We will focus on how to write powerful and easy-to-follow papers in Chapters 5 and 6.

While you spend 30% (or less) of your effort reading and understanding the paper, you should spend 70% (or more) effort thinking creatively and critically:

- **Critical Thinking.** What is wrong with the paper? Are the assumptions reasonable? Is the problem ill-posed? Does the solution have any technical flaws?

  Student A1 read some papers on co-training (an area in machine learning) when he took our graduate course. He implemented some popular co-training algorithms, and tested them on some real-world datasets. He found that the results were not as good as reported in the published papers. This led us (Student A1 and his supervisor) to question: does co-training really work that well? In what situations does it work well, or not so well? After further research, we wrote and published papers in some top conferences and journals in the field. If he did not challenge the previously established views, there would not be these research outcomes.

- **Creative Thinking.** What else can be used to solve the same or broader problem? How can I do it differently, and better?

It is extremely important that you *form a habit of thinking critically and creatively* from now on for whatever you read (even newspapers and magazines) or watch (TV or movies). We always think of "plot holes" when we watch movies (critical thinking), and discuss how to make the movies better (creative thinking)! One of the authors (Ling) also works in child education. One fascinating thinking game he often plays with kids is to read a new story to them or watch a new movie with them. He stops many times during the process and asks kids: what is wrong with the story? How would you continue? How can you make it better? It is crucial to develop a curious, creative, and critical mind, as this is one of the important abilities for doing research (see Chapter 1).

Even when you read a paper published 10 years ago, you could still try to think critically and creatively to "discover" new ideas from the paper. Such discoveries are likely to be re-discoveries (e.g., already published by other researchers). However, if you are used to thinking critically and creatively, when you read recent papers at the frontier of the research in your field, *your re-discoveries will likely be true discoveries.* Those true discoveries will earn you a Ph.D. degree, and help you establish yourself in the field.

We want to emphasize again that in reading research publications, *read less, think more.* There is ample research in education and psychology showing that if people (and kids) are given too much information or instructions, they become less creative and explorative. After all, for researchers, the purpose of reading published papers is to make new discoveries that go beyond the existing works!

Depending on the culture and environment where you grew up and received earlier education, you may already be used to challenging the status quo rather than merely following instructions. If you are not, it may take you some months or even years to change your mentality to be more critical and creative. But if you are aware of it, and try hard, you will improve quickly.

There may be one situation where you do need to read carefully the details of a previously published work: when you need to re-implement it. This may sound paradoxical, as our goal is to do novel research, but it does provide food for thought and ready baselines to compare to, especially when starting empirical research. We will discuss this in detail in Section 4.5.

The second "bad" approach in reading research papers is side-tracking too far and for too long into some minute details, without seeing the big picture. Often research papers contain some technical information and methods that you are not familiar with. For example, a paper on supervised learning may contain a method that uses quadratic optimization as its subroutine, which you may not know well. It would be a bad approach if you get an 800-page textbook on optimization and read it from cover to cover to "prepare" yourself for that research paper.

The right approach is "need-driven learning." Your undergraduate and Master's programs should prepare you with enough basic knowledge for doing research. For the missing knowledge, such as the aforementioned quadratic optimization in papers you are reading, you sidetrack to learn it quickly, and come back and focus on the main papers you are reading.

The key in reading published papers is to know what has been done, and come up with novel ideas. How do you do that?

3.4 WHERE TO GET NEW IDEAS

Scientific discovery, at a very high level, can be regarded as a process of coming up with some new hypotheses, collecting data and performing experiments, and confirming or refuting the hypotheses. But in reality, the process is far more complex. Hypotheses are like new ideas in research. How to come up with new, interesting, and potentially useful hypotheses? Often the hypotheses need to be modified many times while research is being conducted. How can this be done? How to design experiments to convincingly confirm or refute hypotheses? This section discusses how to find new ideas or hypotheses in research.

One risky approach new Ph.D. candidates often use to find new ideas is to read the "future work" section of a paper, or just do the "obvious next thing." However, these are the ideas that the authors have already thought about, and may have been working on, that they are not really the "future" work per se. Thus, there is a chance that if you worked on these ideas, the authors may always be ahead of you. In addition, your mind will be restricted by the authors—you still have not generated new ideas independently.

Earlier, we discussed how you should read less and think more—think critically and creatively. When you are doing that, often you will have questions about the published methods, new ideas and hypotheses about solving the same or different problems, new ways of making assumptions, and so on. You should write down those questions and ideas right away on the margin or the backside of the paper. If you read papers on computers (PDF files), use note-taking software to take notes, or an Adobe PDF Reader equipped with the note-taking function. When you think actively, ideas come and go, and it is crucial to write them down immediately, and study them more deeply later to see if they are indeed new and worth further exploration.

In fact, new ideas can come from other unexpected occasions when you keep on thinking about intriguing research questions. Legend says that the ancient Greek scholar Archimedes had a "Eureka!" moment when he took a public bath and suddenly discovered how to measure the volume of an irregular object (a crown made with gold) by submerging it in the water. He leaped out of a bath, and ran home naked shouting "Eureka!" If you have an "Aha" or "Eureka!" moment when taking a bath, we suggest you also to run to get a pen and paper right away—but remember to always bring pen and paper to the bathroom! Or better yet, bring a laptop or tablet to the bathroom or bathtub, but make sure that it is waterproof!

Sometimes new Ph.D. candidates complain to us that after they read many papers, they find that their "new" ideas have all been discovered! They claim that the more they read, the more difficult the topic seems to be, and the less interesting the topic becomes! This can happen in a well-developed, mature research area. In the next section we will point out that you should pay attention to "hot" and recent research topics when you select your Ph.D. topic. Here, we offer you another very important and effective approach to generating new ideas: having a "brainstorm" session with other people.

A group brainstorming session is a group session where two or more people actively participate and suggest ideas. These people can be fellow graduate students, professors, and other researchers, some of whom may even be outside your research area for fresh ideas. It is actually quite important and useful to develop a "team" of people who help each other during your Ph.D. study. In a brainstorming session, often assisted with amply supplied coffee (or a small amount of beer), *no ideas should be criticized or judged.* New ideas will be drawn and written on a large board, which may help visually stimulate more new ideas. Those ideas can be explored deeply by different graduate students in parallel afterward.

Combining "read less, think more" and brainstorming, a very effective approach we often take is to have one person in the group read a just-published paper first. He or she would present only the problem part to the group, and lead the rest to brainstorm: How would I solve it, and do better? What is wrong with the paper? Again, many ideas and solutions may be found, and can be compared with the published paper. Often new research ideas not mentioned in the paper may be generated, and can be explored afterward.

A crucial aspect of generating new ideas by you or within a group is to try to generate and treasure **bold** new ideas. By "bold," we mean ideas that are radically different from the previous ones, or that refute and overhaul previous work. For example, if you discover that previous work made a wrong or unrealistic assumption and you propose a better one, you probably need to propose new algorithms, and solve a different set of problems. Avoid small, incremental improvements, the so-called "salami" ideas which produce "salami" papers (thin and unsubstantial). A small, incremental improvement is, for example, a different way to search for optimal parameters of the same published work, and by doing so you improve the results by, say, 3%. Your new ideas should almost always be application- or problem-driven, not solution driven. This will improve the impact of your research work.

For example, Student A1 was initially interested in cost-sensitive learning. After reading many papers on this subject, he wrote a survey on it. Writing a survey paper is an excellent way to start researching in an area. He also extended traditional cost-sensitive learning methods, and published a few papers. However, the work was not bold enough to be a Ph.D. thesis. To help the student, we brainstormed: what are other forms of costs in cost-sensitive learning that are useful in real-world applications but have not yet been studied? After several brainstorming sessions, we raised new ideas on different data acquisition costs in cost-sensitive learning. After a literature search, we find out that this extensively studied previously. As this is a bold, new idea in the area of cost-sensitive learning, Student A1 explored the idea deeply, checking into previous works that included the supporting theory, solutions, and application examples. As a consequence of the research, he published several top-rated conference and journal papers, and wrote his Ph.D. thesis on this topic. Since he is one of the first to work on this topic and many people are likely to follow up in this direction, we can say that student A1 "owns" this topic now.

3.5   FROM IDEAS TO RESEARCH AND THESIS TOPIC

A "small" new idea may lead to a conference paper, while several small new ideas on the same topic or one "ground-breaking" new idea can lead to a research and thesis topic. The task of choosing a research and Ph.D. thesis topic is a complex and trial-and-error process. It may also determine your future research direction and career. Section 4.1 provides you with several specific criteria on a good research and thesis topic, proposed by the late computer scientist and Turing Award winner Jim Gray. Here we list some general factors that should be considered carefully:

- Again, **your passion and interests in the topic.** You must be highly passionate and fascinated in the selected topics.
- **Your technical strengths.** Are you theoretical and math oriented? If so you can consider research in theoretical aspects of a field, such as theoretical models. Good at statistics? Consider statistical simulation or optimization, for example. Good at design? Consider a new design and processes. Good at applications? Consider empirical research and killer applications.
- **How new and how "hot" the topic is.** True, one can make novel and significant contributions in virtually any area of science and engineering, but it is often quite hard to do so in 4–5 years working on a Ph.D. thesis if a topic has been studied for many decades, and you could spend a huge amount of time just learning and studying previous works. Thus, you should consider recent topics that are or are becoming hot, and in which rapid progress is being made, as your Ph.D. topic.
- **Your supervisor's vision and opinions.** Your supervisor often has more experience in research and better judgment on new research topics. Sometimes a topic may have been well studied for many years without much breakthrough, even though papers are still being published on the topic; sometimes a topic can be too hard or too easy. Thus, discussion with your supervisor is often crucial. Note, however, as Ph.D. research would advance the state-of-the-art, no one can be 100% sure if a chosen topic will be adequate for Ph.D. study before actual research is carried out.

- **Your future career.** Do you want to be a professor in universities, or a researcher in companies? For the latter case, you should select empirical research where the research project can highlight new ways of applying theories.

When selecting a thesis topic and carrying out research to explore the "unknown territory," is there any way to know "if I am on the right track"?

3.6 HOW DO I KNOW IF I AM ON THE RIGHT TRACK?

Many new Ph.D. candidates often feel in the dark if they are selecting a good topic for their Ph.D. thesis, or if they are doing the right thing in their research. An extremely important way to validate your thesis topic is that *you complete your first iteration of the three main tasks (find new ideas, do solid research, publish papers) early in your Ph.D. study.*

In the later part of the first year, and during the second year of your Ph.D. study, you should start doing research and publishing papers (Tasks 1–3) in research topics you are interested in. Start with some small, interesting research problems, and do solid research (see Chapter 4). If you obtain positive and interesting results, you can write them up as a conference paper, or a short journal paper (see Chapters 5 and 6), and submit it to a conference or journal. You can start with medium competitive (or local) conferences or journals.

As we mentioned earlier, research papers are almost always peer-reviewed by other researchers anonymously. This allows reviewers to write detailed and often critical reviews of your papers. If your paper is accepted, it surely validates your ideas and research topic, and the reviews can help you to further improve your work. If the paper is not accepted, comments from the reviewers will often give you great feedback about your research and the paper. However, it is a bad idea to submit poor and pre mature papers as they can negatively affect your reputation.

Here are some other useful ways to validate your Ph.D. topic and research:

- Discuss your potential topic with your supervisor and other colleagues and Ph.D. students extensively.
- Discuss your ideas with other researchers who work in the same field via emails.
- Chat with other researchers in conferences. Keep in mind that people may give less critical feedback than they might in writing reviews of your papers.
- Find and read recent Ph.D. theses on similar topics from other universities, and compare the depth and scope of your research work to theirs.

To reiterate, it is very important to go through the Tasks 1–3 early in your Ph.D. study. This will not only prepare you for future iterations in your third and fourth year of your Ph.D. study, but also validate your Ph.D. thesis topics.

3.7   SETTING UP A PLAN FOR Ph.D. THESIS EARLY

After you have gone through Tasks 1–3 once or twice, it is highly desirable to set up a "plan" for your Ph.D. thesis, usually near the end of the second year. The plan looks a lot like a two-level outline of your future Ph.D. thesis, with a list of related research problems to be solved. This is usually possible because you have read many recent papers on this topic critically and creatively, and performed some research already.

With the plan, in the next few iterations of your Ph.D. research, you essentially pick up some research problems in the plan, do solid research, and publish high-quality papers. Note however, the plan is not carved in stone—*you need to remain flexible with the plan.* For example, you can skip some research problems if they are too hard or they do not yield enough results. Remember, doing research takes risks in exploring unknown territories. You can also add new, related problems during your research. However, *you must have a strong focus*—all the research problems you will be solving are within a (broad) thesis topic. In the end, your

Cost-Sensitive Learning with Data Acquisition
- **Introduction**
- **Previous work**
- **Cost-sensitive learning with feature acquisition**
    - Sequential strategies [ECML 2005]
    - Simple batch strategies [AAAI 2006]
    - Sequential batch strategy [TKDE 2007]
    - Considering delay cost [new; ECML 2007]
    - Experiments and applications [in all papers above]
    - Real-world deployment [unfinished]
- **Cost-sensitive learning with example acquisition [new]**
    - Partial and full example acquisition [KDD 2008]
- **Theoretical guarantees of acquisition strategies [unfinished]**
- **Conclusions and Future Work**

FIGURE 3.4: A plan of a sample Ph.D. thesis with a central theme and a list of potential problems to be solved, and how it evolves into the final thesis with publications in [. . .]. Some topics are added in later ("new" in the notes), and some were not finished ("unfinished" in the notes) in the final Ph.D. thesis.

thesis plan and the final Ph.D. thesis outline may have an overlap of 50–80%. This way, it is possible that you can make novel and significant contributions and establish yourself in a certain area ("own" an area) at the end of the four-year Ph.D. study, as seen in Figure 3.1.

After Student A1 switched to cost-sensitive learning with data acquisition, we set up a thesis plan together on this new topic. The student worked hard, and published a series of papers in top conferences and journals. This is not surprising as Student A1 has already gone through Tasks 1–3 on earlier research, and can conduct solid research and write papers reasonably well by himself. His final Ph.D. thesis is almost like a simple collection of these relevant papers. See Figure 3.4 for his thesis plan. We put some notes in [. . .] in the plan for papers that were later published. Some topics are added in later ("new" in the notes), but some were not finished in his Ph.D. thesis ("unfinished" in the notes). This plan provided a strong focus in his Ph.D. study, yet he was still flexible with the plan. Note that his final Ph.D. thesis does not include his early published work on co-training. A thesis must be a coherent piece of work, usually on one central topic and theme.

Ph.D. study usually only takes 3–5 years; this is rather short for making significant contributions in research and for establishing yourself in some area.

Having a plan and working systematically (Tasks 1–3 in Chapter 1) according to the plan *is crucial to make this happen.*

3.8   LEARNING TO ORGANIZE PAPERS AND IDEAS WELL

A related but important issue in undertaking the long and arduous task of completing Ph.D. research is the need of a good system to organize the many papers that you have read, notes on ideas you have thought about, and references you have used and cited in your papers. Otherwise it can be hard to find those things during research and when you write your thesis. In the "old" days when we worked on our theses, we created piles of physical papers on different topics, and used text documents to keep track of papers, ideas, and so on. Nowadays most papers are in electronic format, such as PDF, or portable document format, created by Adobe Systems, and there are quite a few "reference management software" that can help you with all of these tasks with ease. Two popular ones are Mendeley[5] and Papers.[6] Also, you can use ebook software, including iTunes and Kindle to organize PDF files.

You can create your own note-taking system, or choose a software package to organize papers, notes, and references, and you need to use it effectively to manage your research the in several years of your Ph.D. study.

·   ·   ·   ·

[5] http://www.mendeley.com/
[6] http://www.mekentosj.com/papers/

CHAPTER 4

Conducting Solid Research

In this chapter, we go deeper in describing the process of conducting solid research. We start by giving a high-level overview on the research process (Section 4.1). We then discuss some high-level criteria, called Gray's Criteria, to judge what research ideas and projects are worthwhile to pursue (Section 4.2). Subsequently, we discuss how to form hypotheses to work on in order to pass the criteria (Section 4.3) and how to confirm or dispel the formulated hypotheses (Section 4.4). We further discuss some other objectives as you go through your research process, such as how to brand yourself (Section 4.5), how to choose between theoretical and empirical research (Section 4.6), and what to expect in team work and multidisciplinary research (Section 4.7).

4.1 AN OVERVIEW OF A RESEARCH PROCESS

We refer to the steps in which we conduct research as a research process. A research process, broadly speaking, consists of the following phases:

Step 1: Find a novel and potentially high-impact problem to solve based on the high-level criteria (Section 4.2) on choosing a good problem.

Step 2: Find a candidate solution. You then formulate a hypothesis as a (problem, solution) pair. Each hypothesis states that the solution can indeed solve the problem.

Step 3: Conduct theoretical or empirical analysis on the hypotheses to confirm or dispel them. You also conduct significance tests to establish the validity of your conclusions on these hypotheses.

Step 4: If you are successful in your significance testing, then you are done. Otherwise, you return to Step 2, find another candidate solution to the problem, and repeat the process.

Your initial solution to the problem might be a simple one, and there is a good chance that you will fail to solve the problem either because the problem is too hard to solve, or because the solution does not solve the problem. The reason that you often refute your initial solution is that the solution is usually very simple to start with, and other researchers working in the same field may have already thought about the solution to the problem, tried it, and refuted the hypothesis. However, as most published articles only report on "positive" results, you might not be aware of the fact that the solution in fact is not a good fit for the problem, and thought otherwise!

In research, we almost always have to revise and "tweak" our initial hypotheses to "make them work," so to speak. Finding the right problems, and forming and revising the most appropriate solutions for these problems, is a highly complex and creative process, and thus it is hard to "quantify" this creative process precisely. We will describe a so-called *Research Matrix Method*, which inspires us to make *systematic* associations between problems and solutions. This often leads to new hypotheses and novel ways of solving the same or different problems. See Section 4.3 for detail.

How do we confirm that a hypothesis is true? There is a very broad spectrum of approaches. On one end of the spectrum is a (pure) theoretical approach. It uses mathematics to represent the hypothesis, and logical reasoning and mathematical derivation to prove (or disapprove) it. However, often assumptions need to be made so that proofs can be carried out. One of the most important and classical theoretical work is Euclid's *Elements*, written in 300 BC. It starts with definitions and some basic assumptions (called axioms or postulates, such as two parallel lines will not intersect if they are extended in any direction), and uses logical reasoning to prove (confirm) hundreds of propositions and theorems, including the Pythagorean Theorem. When a theorem is proved once, it is true forever. There is no need to "run experiments" to verify proven theorems. The key questions to answer are how strong the assumptions are, and how the results can be applied in practice.

At the other end of the spectrum is (pure) empirical research. Often such empirical research is closely related to practice, as empirical methods can be easily turned to applications. Thus, do not be afraid to pose some bold hypotheses that represent *killer applications*—solutions to certain problems that may bring direct benefits to a large number of people. Empirical hypotheses testing can often be stated in a natural language, say English, and can only be confirmed by running experiments on simulated or real data. Experiments can be conducted with people, animals or plants, materials and objects. They can also involve computer simulations, and computer programs taking various datasets as inputs. When conducting these experiments, it is crucial that researchers have a sharp eye for interesting and anomalous results. This is another important way to find surprising or new hypotheses during research in addition to the research matrix method we mention in Section 4.3.

We often use different materials, datasets, and have different experimental settings in experiments. Do we need to exhaust all possible materials, datasets, or experimental settings, to show our new design and methods are better than previously published results? What if this is impossible? How can we compare with previous design and methods published by other people that we do not have access to? Your decisions on these issues will determine *how thorough and rigorous your experiments are to support the validity of the hypothesis*. The degree of such empirical evidence with experiments can be quantified, fortunately, by the *Significance Tests*, which is a subject in statistics. Here the term "significance" does not mean importance or impact, as we have been using in this book. Briefly, a significance test is a statistical procedure to test if a conclusion (e.g., your method is better than previous ones) is likely to happen by chance. See Section 4.4 for more details. In order to pass the test, repeated experiments and head-to-head comparisons usually need to be conducted as part of your research.

A more rigorous support can be obtained if you can ask authors of the published papers to apply their own methods on your materials and data, and your new method is still shown to be better than theirs. This is because the original authors can usually apply their own methods better than you could. However, if you only compare your method with those results in some published papers (without reimplementation of these methods on your own), the evidence that supports your hypothesis is considered weak, as the experimental settings can be different, and you might not satisfy the assumptions needed to conduct the significance tests. See Section 4.4 for more details on this.

4.2   JIM GRAY'S CRITERIA[1]

Many students think of research as an admirable career, where one can truly make a name in history. Just consider examples in our textbooks: Newton, Galileo, and Einstein, to name a few. However, for a beginning student, research should be both a curious adventure and a serious business to manage. Let's start with the fun and adventurous part first.

One of the authors (Yang) once attended a lecture given by Professor Charles Townes, an American Nobel Prize-winning physicist who made major advances in quantum electronics for lasers and masers. Yang had earlier met Townes when he freshly graduated from Peking University when Townes first visited China on a science mission as President Jimmy Carter's representative in 1982. In this later seminar, Townes reflected on his joy in his very first discovery, a fish. He said that in his childhood he used to catch fish. One day he got a fish that he could not name according to an encyclopedia. Curious, he wrote a letter to the Smithsonian Institute, a US museum. He was pleasantly surprised when he got a letter in return, noting: "Congratulations, Mr. Townes, you have discovered a new species of fish!" Imagine the joy and excitement this boy experienced when reading these words!

---

[1]Jim Gray was an American computer scientist who was credited for major developments in database theory and a proponent of the fourth paradigm in scientific discovery.

Discovery is a childhood joy. As we note in our previous chapters, one enters the research field primarily because innovation and discovery are fun. It is a dream we all have had since childhood. What we want to say in this chapter is that to be prepared for a career in research, we must have more than curiosity; we must have a process in which we can manage our research like managing a business. Just like running business, if we happen to manage it well, it will flourish and prosper. Our impact in the world and our contribution to knowledge will be significant. Otherwise, if we do not manage it well, we will find ourselves frustrated and discouraged.

Once we know that we are motivated to do research, our first task is to decide on our objectives and directions in which we will go deeper. This process will definitely be curiosity driven, but should also follow a well-managed process. As noted in previous chapters, finding a good topic takes many activities, ranging from talking to experts in the field to critiquing many papers and works in your field of interest. But if there are several areas that you are more or less equally interested in, which one should you choose? How do successful people pick an area to work on? Why are some equally talented people not as successful and impactful as others?

In our careers, we have been talking and listening to many "successful" researchers. One of the best-known criteria for high-impact research was summarized by the late great computer scientist and Turing Award winner Jim Gray, who was credited with the invention of modern database technology and the fourth paradigm in computer-science research. In his 1999 Turing Award lecture, he stated that good research should

- have clear benefit,
- be simple to state,
- have no obvious solution,
- have a criterion where progress and solutions are testable, and where it is possible to break the larger problem down into smaller steps, so that one can see intermediate progress after each step.

We will refer to this set of rules as "Gray's Criteria" hereafter.

Take word processing as an example. Suppose a student is very interested in proposing to do research in word-processing technology. This technology has clear benefit, which is very simple to state. However, it is obvious today how to do it, via a variety of word-processing software. Thus, the research is not impactful, unless it is a dramatic improvement on the existing word-processing software.

However, suppose that the student has proposed to invent a new word-processing system based on speech input instead. This has clear benefit for a variety of people, including the handicapped. It is very simple to state, and it is not yet obvious how to do it accurately since in the speech recognition field today, we are still far from having a robust recognition technology based on voice input. The progress can be testable: by having humans speak to a microphone and producing coherent and accurate text. Finally, the objective can be broken down to smaller steps, including phoneme recognition, sentence recognition, grammatical analysis, and fault tolerant, etc. Thus, this can be considered a good direction to pursue for the student.

Going back to our earlier example of Students A, B, and C. Recall that **Student A** is a type of student mostly interested in fundamental and academic research. Earlier we described A1 in detail. Let us use A2 to denote another student of this type. For Student A2 to find a research problem, the question becomes: how do I advance the state-of-the-art in academic research? After reading research papers and proceedings, Student A2 found the topic of "transfer learning" to be a very attractive topic, which studies how knowledge gained by an agent in one domain of expertise can be usefully transformed into knowledge in another, related domain. Transfer learning tries to endow machines with the ability to learn from its experiences that are drawn not only from a single domain of interest, such as examples of categorizing images, but also from other different but related domains of interest, such as reading books. The topic of transfer learning also has its psychological and educational foundations, as it is an integral part of human intelligence. People have the ability to carry out transfer learning, for example, when they learn to plan military and business strategies via their knowledge in playing chess.

Student A2 found the topic of transfer learning intriguing enough to continue exploration. In particular, he found the topic of transferring learned models between text and images very interesting, and he started to conduct a thorough survey of existing works in transfer learning, and then applied the **Gray's Criteria** to the topic.

The topic has clear benefit to many computer science areas, such as data mining and machine learning applications, as these algorithms are a major component of search engines, social networking applications, and business applications. However, a major bottleneck in these applications is the lack of high quality annotated data, because data labeling is a time-consuming task often done by humans. Transfer learning allows one to borrow the knowledge gained from other fields to alleviate the dependency on annotated data in a given domain. In Student A2's problem context, transferring knowledge gained from text documents to images allows an online social media system to categorize and search for images of interest much faster and more accurately than before.

The problem is also simple to state. Try to tell an ordinary computer user that transfer learning allows a search engine to adapt to their personal preferences much better than before, or tell a credit card user that card misuse and fraud detection can be made more accurate, and you will know why. In the case of Student A2's problem, a story of teaching computers to recognize images by reading text sounds very fascinating!

While we humans apply transfer learning every day in our lives, it is in fact not that obvious how to do it on a computer. Previous approaches in machine learning and data mining relied on the assumption that the training and testing data are represented similarly. This assumption is no longer valid in Student A2's problem.

To Student A2, progress and solutions in transferring from text to images is clearly testable, as we can use the percentage of correctly categorized images as a measure of success. This measure is a function of the number of text documents fed to the machine for reading, which can be displayed clearly as a chart.

Finally, the problem can also be broken down into smaller steps, so that one can see intermediate progress. For example, Student A2 can try to work with an existing translator from text to images. Then he can design algorithms and models to learn the translator from data. Finally, he can consider varying the types of text and images and observe the performance of the resulting transfer-learning system.

Now let us turn to another student of type B (empirical research), Student B2, who is interested in large-scale application-oriented problems, with an aim to join research labs in industrial companies after graduation. For Student B2, the transfer-learning problem itself is interesting, but he is interested in seeing large-scale applications at work. To do this, he found an application idea in the area of query classification in search engine design, where the problem is to automatically categorize user queries into categories. For example, when we type in a query "hotel" in a search engine, the system must deduce that the user might be interested in a commercial purchase, thus providing the user with informational results about hotels, as well as advertisements. Making transfer learning work in this search context allows a search engine to quickly adapt to changes in a commercial search and advertising application. This would allow new areas to be quickly developed, and search-query categorization can be readily extended by "transferring" previously annotated query data to new application domains. To demonstrate the innovation, Student B2 plans to build a large-scale data mining system based on transfer learning based on data collected by a search-engine company. To make the data access feasible, he decides to join a commercial search company for one semester as a student intern.

For a student of type C (application and entrepreneurship), Student C2, whose aim is to become an entrepreneur, may decide to focus on a more profitable area, online advertising, in order to gain more experience in the area of business intelligence to prepare for his future company. Advertising is a science as much as an art, which touches not only on business as a topic, but also economic theory, game theory, human-computer interaction, artificial intelligence, and data mining. Since carrying out this research involves a multitude of disciplines, the student decides to go to various academic as well as business conferences to enlarge his social network. He goes to some more focused and commercial workshops more often, and plays a major role in many functions in conference or workshop organizations such as being a student volunteer.

One of the most important aspects of Gray's Criteria is that the progress should be **testable**. Today, much research is done by means of system building and empirical testing. In this context, a testable criterion can be further refined to: we must ensure that it is feasible to get credible and a sufficient amount of data. In many student cases under our direct or indirect supervision, we have observed that this is a critical point where many students get stuck: they have invented great ideas, algorithms, and theorems, and implemented systems. The only thing that remained to be resolved is credible verification, and this is where the students often fail: they cannot find credible datasets beyond publicly available toy-ish data, which breaks down the whole process of research. For example, in computer science and engineering, some students conduct research by proposing new algorithms that rely on search-engine log data. However, this data cannot be accessible at university labs. Some of the data are only available via industrial companies such as search-engine companies. In this aspect, students should seek out opportunities to work in collaboration with industrial research labs, where they can take on the position of a research intern, to complete their empirical research. Indeed, this has become one of the most effective ways for students to carry out their research plans.

4.3   THE RESEARCH MATRIX METHOD

Gray's last rule says that research should be able to be broken down into smaller steps, so that one can see intermediate progress. This is important, since as we know well in history, new innovations are possible by "standing on the shoulders of the giants." This applies to great inventions. It also applies to your own research agenda, in which one of these giants may, in fact, be you.

One of the most effective methods is the so-called research matrix method, in which one can systematically plot a path to successful research. In Figure 4.1, we see a matrix partitioned via an x-axis, which represents different methods and techniques that we have and a y-axis where we list the potential problems and subproblems we try to solve. Some of the subproblems are related and should be solved in sequence, while others can be solved concurrently.

To make the methodology more concrete, let us consider the case of a student who is interested in machine learning and data mining areas, as we mentioned before. Suppose that the student is interested in making a career in cost-sensitive learning. Cost-sensitive learning concerns how to build a computer-based model that can improve itself over time by considering different costs associated with different types of mistakes made by the model on future examples. This area is particularly important because many practical machine-learning problems are intrinsically cost-sensitive, in that the data is unbalanced. A good starting point is to carry out an extensive survey of this area, as we noted in Chapter 2. Suppose that the student has broken down the cost-sensitive learning area into several sub-

The Matrix Method

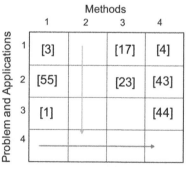

FIGURE 4.1: The research matrix method.

problems: cost-sensitive learning when the costs are known, cost-sensitive learning when costs are unknown ahead of time, cost-sensitive learning when the cost is associated with missing data acquisition, and cost-sensitive learning when data comes in stream form. On the method dimension, the y-axis, one can lay out such methods as Bayesian classification, ensemble methods, density-ratio methods, and partition or tree-based methods. Furthermore, one can list out various online learning methods.

By examining this matrix, the student can first try to fill in various cells by papers he or she has read before: each citation index such as [3], etc., represents one such paper. The student can optionally put notes in each cell, to denote the strengths and weaknesses of each paper upon reading. The student can find various surveys and comments on the Internet via search engines, which can get the student quickly into focused areas. These are the areas where there is no existing literature, or few literature pieces, to occupy a cell. Examples in the figure include the column #2, row #4, or cell #(3,3).

First, cell #(3,3) offers a perfect opportunity for the student to enter the field, or "get his or her feet wet," so to speak. The student can try to apply method #3 to subproblem #3 in order to get the necessary practice. Others in the machine-learning field will appreciate the relatively significant impact because this proposed work clearly offers a novel solution for an important subproblem.

If this is truly the case, as this is how many of us more senior researchers often started our research career, congratulate yourself—you have entered the door of research. This is where you can taste the sweet fruit of your own discovery for the first time. However, more exciting things await you in the matrix, so read on!

Example: Student A2 is most interested in pursuing an academic career. He has found a topic in the area of "transfer learning," and sets out to find a suitable topic in this area for his Ph.D. research. After reading the relevant journals, proceedings and articles, Student A2 lists the specific transfer learning related problems as "knowledge transferring in the same feature space," "knowledge transferring across different feature spaces," "knowledge transferring with unbalanced data," and "knowledge transferring from multiple source domains." On the method dimension, he lists: "instance-based transfer," "feature based transfer," "transferring with ensemble of models," and "transferring across multiple learning tasks." From the matrix, he noticed that there has been considerable previous works by other researchers on "knowledge transferring in the same feature space," "instance based transfer," "knowledge transferring across different feature spaces," "instance based transfer," etc., but there is relatively little work under "transferring across different feature spaces," "transferring across multiple tasks." After discussing these findings with his supervisor and supervisory committee members, he decides that this is an area to pursue. He then writes a proposal, formulates several plausible solutions, and sets out to find appropriate data for the experiments. Finally, Student A2 gets several papers published and completes his PH.D. work.

In contrast with Student A2, Student C2 is more interested in an industrial and entrepreneurial career. Thus, Student C2 decides on a different topic to pursue, but he can still make use of the matrix method. For example, Student C2 finds that the topic of recommendation systems (such as that used by Amazon.com) to be significant and of practical importance. In this area, he finds that the problems can be formulated as a vector "recommendation based on dense matrices," "recommendation based on sparse matrices," "recommendation with an external ontology for users and products," and "evolutionary recommendation with temporal information." On the solution dimension, he lists "user based recommendation," "item-based recommendation," "model-based recommendation." This gives a matrix where Student C2 starts performing a literature search and filling in the cells with references. Student C2 soon finds out that the area of "evolutionary recommendation with temporal information," "model-based recommendation" is both interesting, new and will be of tremendous commercial value if introduced to some well-known commercial online-shopping companies. This idea is confirmed by his committee members, which includes a researcher from a well-known industrial research lab. He sets out to do his research, which can be tested on large-scale data and using well-established evaluation criteria. With this work he applies for a patent and then writes a business plan for a university-based spin-off company.

4.4   CARRYING OUT YOUR RESEARCH

Once you have set the research objective and have done the related literature survey, you are ready to conduct your research. A first important step is to state your research objective in one understandable sentence. For example, if your objective is to show that you can design a better search engine than Google, then you might want to be more specific about what you mean by "better," and formulate your hypothesis more concretely. For example, you might state "my hypothesis is to use social networking information in Web search results toward more accurate ranking results compared to not using such information in Web search." This hypothesis can be easily understood by even non-computer science people, and can be concretely confirmed or disproved. More concretely, the hypothesis can be further instantiated by the following statement: "my algorithm ABC uses social networking information of masses of users as well as hyperlink information. We expect that ABC gives better Webpage ranking results than Google's PageRank algorithm[2] when the former uses additional social networking knowledge."

It is more advantageous to make your hypothesis more concrete. First, having a more concrete hypothesis at the outset allows researchers to focus on a few key elements in your work. For example, it allows you to formulate your methodology and present your ABC algorithm as clearly as possible, so that others can repeat your experiments and confirm your hypothesis. Second, it allows you to decide on the necessary baselines and evaluation metrics in order to confirm or disprove your hypothesis. For example, you may formulate your methodology as one that is based on a theoretical proof that no matter what PageRank algorithm provides in the final result ranking, your ABC algorithm is going to perform at least as well as that by PageRank. In order to show this, you may realize that you need a widely accepted evaluation metric in order to compare the ranking results.

[2] http://infolab.stanford.edu/~backrub/google.html

For example, you might adopt the "Area Under the ROC Curve" (or the AUC measure) as a ranking metric.[3] Alternatively, you may wish to use a metric known as Normalized Discounted Cumulative Gain (NDCG[4]). Whichever metric you might choose to use, you need to follow up with a proof that your ABC algorithm will outperform the PageRank algorithm by a significant margin. If you are more theoretically oriented, you might look for proof methods that can help you show that your hypothesis holds with a series of theorems and corollaries.

In many cases, however, the hypothesis can only be answered empirically. If you adopt this path, you'll need to design the experiments, which involves several key steps. First, you'll need to identify the datasets and sample the necessary data for your experiments. Often the original dataset is noisy, incomplete, and large, and for this reason they are often called the raw dataset. For example, for showing the superiority of a new search method, one often needs to obtain large quantities of search log data from a real search engine. This raw data is often extremely large and incomplete, but a sample of the data that is properly cleaned can be used to represent the general real world situations that are encountered by a real Web search algorithm. In this case, a section of the data that involves several months that cover the four seasons of a year might suffice. When there are multiple segments of the data that vary in distributions, it is better to sample each part independently. Furthermore, it is a good idea to apply stratification by partitioning the entire large population into small segments, where the elements of each segment are sampled out randomly. The sampled data is expected to resemble the distribution of the original dataset. Furthermore, when the data have much variance due to unusually high uncertainty, it is a good idea to apply Monte Carlo methods for sampling, where the basic idea is to run repeated computer simulation to generate the data according to the input parameters underlying the data distribution.

---

[3]J. Huang and C.X. Ling. 'Using AUC and accuracy in evaluating learning algorithms.' In IEEE Transactions on Knowledge and Data Engineering, volume 17. 2005
[4]K. Jarvelin, J. Kekalainen: Cumulated gain-based evaluation of IR techniques. ACM Transactions on Information Systems 20(4), 422–446 (2002).

An important aspect of experimental design is to decide the free parameters or independent variables that you must deal with, so that you can decide on their effect on the dependent variable (the search ranking result in our Web-search example). It is important that one of these variables is allowed to vary, while others (known as control variables) are kept at constant values or randomized. For example, in the search engine design problem, variables include the number of Web pages, the in-degree and out-degree of hyperlinks of each page, the length of each page, and the number of topics discussed in the content of the pages. When all of these parameters are fixed, we have a scenario that represents a snapshot of the real world. When we vary one parameter while holding all others constant, we create a performance comparison chart on which we display both the performance of ABC and PageRank, in order to ascertain the superiority of ABC as a better method. This experiment is often repeated by varying one variable at a time for creating a collection of charted results.

Often, even after the experimental metric and comparison baselines are settled, we still have to decide on how to obtain credible subjects on which we obtain the results. For example, in the Web search case, we might hire several users to judge the quality of ranking results in response to a search query. Here, we have to decide on a collection of benchmark search queries that other researchers also use in evaluating their algorithms. We must also ensure that the users have a sufficient degree of independence and diversity to ensure the generality of the result. In fact, there is a large amount of literature in psychology on how to choose test subjects to ensure the most credible results.

When dealing with models that can be constructed based on the data, it is important that we do not mix the training data used for model building with the data for testing and validating our models and hypothesis. In experimental design, this is known as hold-out tests, in which we hold out a portion of the data purely for evaluating the obtained model. When the data are in short supply, we can also use *N-fold* cross validation in which we partition the data in N folds, and let each fold take a turn in being the test data with the rest as the training data. This results in N results that can be averaged to produce the final, more reliable result. In doing this, an important issue is to show that the comparison results are consistent under different experimental conditions, via statistical measures such as variance and confidence intervals.

Finally, once we complete the experiments and produce the results, say in the form of a number of data tables, we must follow up with interpretations and analysis of the results. There are two aspects here. First, we must be able to tell the reader whether the experimental results from competing methods showed significant differences via statistical significance tests. Second, we must explain the significance of the results via what we know from the domain knowledge; in our search example, for example, we need to explain what a certain difference in ranking result would translate to in user experience and search quality.

Now we return to the topic of conducting statistical significance tests on your results. The basic idea here is to ensure that the results are not due to random chance. This can be done by setting a significance level at a low value, such as 5%. Then, we can establish that the probability that result is by chance is lower than 5%, and subsequently, we can say that our result is significant; in this case, we can reject the null hypothesis that there is no difference.

The significance testing can be done via various significance tests. The student's t-test, for example, tells if the means of two sample groups show significant differences. It starts with a null hypothesis that there is really no difference to be detected between the means of the two methods, and set out to show that the null hypothesis can be rejected. The student t-test is used to determine whether there is a difference between two independent sample groups and when the sizes of the groups are reasonably small. Another often-used significance metric is p-value, which is the probability of observing the test statistic value if we assume that the null hypothesis to be true. We often reject the null hypothesis when the test statistic is less than the significance level α.

Summarizing, the research process starts with the formulation of one or more hypotheses (often a null hypothesis and alternative hypotheses), and goes through a cycle in which an experimental design is completed together with data selected and sampled, and then the statistical significance of the results ascertained. Finally, domain-related explanations are given so that others can fully appreciate the impact of the findings. In the process, it is important that the experiments are repeatable by others under the same experimental conditions.

4.5   BRAND YOURSELF

Taking the view of the matrix (again, see Figure 4.1), scientific research itself is not very different from what commercial market researchers typically do when launching a product or a company: they have to find out where the market needs truly are.

On examination of the matrix again, a more important discovery from the matrix is that an entire area can be "owned" by the young researcher, through which the researcher can truly brand himself or herself as an expert in an area with a significant impact.

Let us go horizontal first (see the horizontal arrow in Figure 4.1). This is when you have a good grasp on a significant and high-impact subproblem, perhaps due to your fastidious experimentation and examination of a particular problem at hand, or perhaps because you have gained an important insight by reading many research papers and noticed a lack of attention to a hidden gem. Once you have a good understanding of the nature and complexity of the subproblem, you start asking the questions: "Can method [x] be applied to this problem?" "What impact will be brought forth if I succeed in applying method [x] to this problem?" and, "if method [x] cannot be applied to this problem, what knowledge is gained in the field of research?"

When going horizontal in this methodology, is it critical to compare the different methods [x] and [y] on the same problem to elicit their relative merits? "What are the advantages and weaknesses of method [x] as compared to method [y]?" It may be the case that method [x] is more efficient than method [y], whereas method [y] is more accurate, by some measure of accuracy, than method [x]. It might also be true that method [x] is better in reducing the error rate, if we continue with the cost-sensitive learning example, or that method [y] performs better in some other measure of success such as AUC in ranking. In other words, many exciting research questions can be raised and answered by conducting this research horizontally, and as a result, many top-quality research papers can be written.

Likewise, we can also go vertically from top-down in this matrix (see the vertical arrow in Figure 4.1), by considering applying the same method to a variety of related problems in one focused area. This is preferred if you have a very good grasp on the method's inner strengths and weaknesses. As an example, suppose we choose the decision-tree learning method in machine learning, and wish to apply this solution to various subproblems of cost-sensitive learning. Then we may ask whether the method performs as expected on unbalanced class classification problems, on cost-sensitive learning problems when we have no good knowledge on the cost function, and on cost-sensitive learning when human feedback (which are also costly) are considered in the learning process for future data. We can compare the decision-tree based methods to all other methods that others have tried before on this problem, and if possible, we can find out why the decision-tree based method can perform better or worse than these methods. In this way, we can find out the method's true merits and weaknesses in various learning problems that involve costs.

Like the horizontal research methodology, this top-down process also allows you to truly *brand yourself* in your research community, and make a name for yourself. A popular analogy in many research circles is that a research method is like a hammer, whereas a research problem is like a nail. In this analogy, the horizontal methodology is akin to finding different hammers for the same nail, where as the vertical methodology is like finding different nails to hammer on by the same hammer. As the saying goes, "once you have a hammer, everything looks like a nail!"

Finally, a better solution is if you can identify both a hammer and a nail, by going both horizontally and vertically (e.g., a submatrix from cell# (1,1) to

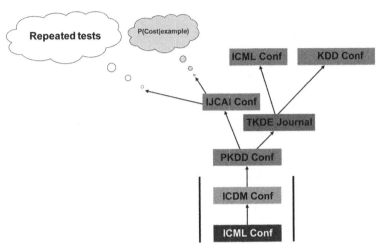

FIGURE 4.2: A tree of research results.

cell# (3,3) in Figure 4.1). If you can truly practice this "hybrid methodology," you may already have a research career at hand, and a tree of fruits will grow with the associated publications, as demonstrated in Figure 4.2.

4.6   EMPIRICAL VS. THEORETICAL RESEARCH

The above discussion naturally brings us to a popular question raised by many students: should I do theoretical research or empirical research? In engineering, the corresponding question is: should I do fundamental research or application-oriented research?

As we discussed so far, the researchers' job is to uncover the causal relations between factors that may be hidden in nature, and this job can be accomplished either theoretically or experimentally. Generally speaking, theoretical research fits better for areas where well-founded assumptions can be formulated to allow inferences to be drawn from, to verify one or more phenomena under observation. In contrast, empirical research is when such assumptions are difficult to come up with, thus requiring experimental verification and data collection. In traditional science such as physics, the boundary between theoretical and empirical research was clearer. Nowadays, especially in engineering science and practice, the line between theory and experimental research is less clear. Add to this gray area the term "system oriented research," which may indeed confuse some beginners and even seasoned professors. We have more than once heard some senior professors and researchers proclaim: "we have built this system, which is my research."

We wish to clarify, however, in this section, that building a system without subsequent analysis is not research, yet. But, if you have a hypothesis, say "Method X is better than Method Y under conditions Z," this hypothesis can be verified by building systems to realize X, Y under Z and evaluate the hypothesis with these systems. This way of proving or disproving a hypothesis is research.

Let's start with empirical research, which is often associated with concepts such as system building, gaining experiences, and making observations. Often, empirical research starts with a hypothesis, which is a prediction or proclamation about the world. As an example, the claim that "it is possible to fly faster than the speed of sound" is a hypothesis. It can be proved, or disproved by building a plane that can break the sound barrier, which provides evidence that is observable.

In science and engineering, hypotheses and empirical research are often tightly coupled, but they may not be raised and verified by the same person or the same group of researchers. Theoretical research proposes models and hypotheses that can be generalized and tested. They often state a fact that goes beyond the current set of observations using methods of induction. Once the models are built, theories should have the ability to deduce new facts about the future. Thus, good theories have the generalization ability. Empirical and theoretical research also often go hand in hand; in fact, in physics, the theoretical and empirical research works are often done by theoretical physicists and experimental physicists, respectively, who are often two separate groups of people.

How do we get the initial ideas in empirical research? One often-practiced first step is learning by repeating some other researcher's work and rebuilding their systems. For example, if you decided to build a search engine system for scientific literature, then perhaps a good system to try to build is the PageRank algorithm that underlies the Google search engine. There are several advantages in doing this. First, to get one's hands dirty and to understand the details of system building is very important in system and engineering oriented research. By repeating others' work in your own environment, you will understand what is stated in their research papers much more deeply, and even be able to understand what was left out of their papers; in some cases, many important details are left out either unintentionally or because of space limitation. However, unless you try building the systems on your own, you will not be able to fully understand these details. Second, building these systems allows you to appreciate the complexity and scope of the problems at hand, and think hard on how to plan for your own research in the future. Often, one cannot understand fully the issues and resources that are involved in completing a project. It might be the case that, just by reading about a system in a research paper, one often underestimates the scope of the project. This is particularly troublesome for beginners, as a major delay in system building can disrupt the long-term plan for one's Ph.D. study. Third, building a system allows you to have an important resource at hand: the baseline or benchmark system, against which you can later compare your more superior system to and demonstrate the validity of your future hypotheses. Finally, and perhaps most importantly, going through an entire process of building a system allows you to reveal many of the weaknesses or oversights of a prior work, and allows you to propose your own hypothesis and formulate an innovative idea. Back to our earlier search-engine example: you might find that the original PageRank algorithm did not take into account the individual users' preferences and interests, and decide to develop a personalized search engine for scientific literature in response.

To see whether you as a researcher should become a theorist or an experimental scientist, consider that in science and engineering, the development of theories have been serving two strikingly different purposes. The first purpose is for experiments to confirm or disprove a preceding theory, and the second one is vice versa.

In the first scenario, on the one hand, theory comes first, which should intrigue a sufficiently large number of subsequent researchers, who then build systems and conduct experiments to prove or disprove the proposed theory.

One example of this mode of research is Artificial Intelligence. In his famous article, "Computing Machinery and Intelligence," Alan Turing put forward a grand hypothesis for all to see: that digital computers built based on Turing machine can be made intelligent on par with humans. This hypothesis cannot be proven or disproven by theory only, just like the hypothesis that the Earth goes around the Sun cannot be verified based on theory alone. One should actually build an intelligent computer to pass the Turing test to be claimed intelligent, and this is both systems work and experimental research. The theory is testable and is also decomposable, thus we have various branches of AI that includes computer vision, machine learning, planning, game theory and computer games, etc. This is an example of theory before experiments.

On the other hand, we can have experiments before theory in many other examples. This used to be largely how physical science was done. You do experiments, and produce results that do not connect with, or cannot be explained by, a current theory. You now have something about which to posit a theory. The Michelson–Morley experiment performed in 1887 created a "crisis" for classical physics, and then Einstein proposed Special Relativity (a new theory) that can explain the experiment outcome. Here is a more recent example in the development of modern machine learning theories. In the early 1980s, there was great excitement about the fact that digital computers could be made to learn from examples. Many schools of learning algorithms and systems flourished, each backed by a number of successful applications. Noticing this diverse development in the computational learning field, and the inability of existing theories to unify the seemingly different learning algorithms, Leslie Valiant from Harvard University developed the PAC learning theory, which stands for probably approximately correct learning. According to this theory, learning is simplified as the ability for a system to make fewer and fewer mistakes with higher and higher likelihood, given that more examples are input. Several assumptions are made on how the examples are drawn, how the mistakes are measured, and how to define when the learning is successful. In most of the real-world machine learning applications, these assumptions are not necessarily true, or cannot be easily verified. Nevertheless, the theory was abstract, elegant, and simple, and very useful in explicating the differences between different learning algorithms. For this and other works, Valiant obtained the 2011 Turing Award.

Are you more like Alan Turing or Leslie Valiant? The answer is that most likely you will need both abilities in your work. As a beginning researcher, you need to have a keen eye on the way the general field is going, and able to see the theoretical underpinning in the trend. At the same time, you need to cultivate a keen eye on how to unify different empirical research and various theories with a simple and elegant theory of your own. This may happen in a computer science thesis on machine learning, where one can propose a new learning problem, offer a solution, experimentally compare different approaches, and develop a theory to prove the algorithm's convergence and efficiency, under certain assumptions. In other words, the world has become more complex that one had better become both a theoretician and an experimental scientist.

Now we return to the question of the value of building systems. As stated earlier, one cannot claim that one is a researcher by way of building systems only. However, one can become a researcher while building a system with a hypothesis in mind. The fact that you build an airplane is to prove that heavier-than-air machines are able to fly in air. Likewise, the fact that we build a computer system is to prove that the machine can be made to accomplish one or more of the intelligent tasks that, before the system was built, humans were still better at these tasks. An example is the IBM's WATSON system, which beats human champions in the TV game of Jeopardy, which before WATSON was only achieved with smart humans. Thus, before building your system, ask yourself: what are you trying to prove?

In our example of Students A, B, and C, the difference between theoretical versus empirical research can be reflected in the choices of A and B, who prefer theoretical and empirical research, respectively. Let a particular instance of Student A be Student A2. Student A2's research focus is in how to learn a better model by learning from examples that can come from very different source domains, where these other domains may use a different feature representation. Student A2 further decides to use a method of ensemble model learning as a solution. To proceed, he first asks the question: "is it possible to transfer useful knowledge between any pair of sources and target learning domains?" To answer this question, he consults the theoretical works in computer science and mathematics, and formulates a set of sufficient and necessary conditions with theoretical proof of transferability. This set of theorems then guides his search for a better empirical solution, which he then verifies further with large data sets.

In contrast, Student B2 (an instance of Student B) prefers to follow an empirical approach from the beginning in the research area of knowledge transfer. To do this, he finds a specific application domain of online product recommendations, and realizes that the main challenge in this area is to come up with a scalable set of algorithms that can work both accurately and efficiently. He then consults research papers in this area, which he extends into an innovative algorithm. He tests the algorithm on several large data sets obtained from online product recommendation and advertising companies, on which he realizes good performance.

4.7 TEAM WORK AND MULTI-DISCIPLINARY RESEARCH

A long, long time ago, research used to be a one-person affair. Leonardo da Vinci studied animal and human anatomy, architecture, and physics alone. Galileo Galilei studied the motion of objects and planets, Newton published his three laws of physics, and they published their works all by themselves. This situation has dramatically changed today: open any issue of *Nature*, and you will notice many co-authors for each paper; sometimes the authorship takes up an entire page. In milder cases, look up any author in engineering science, in Google scholar or DBLP, and you will find an active researcher who has likely co-authored several, if not many, research papers with dozens of other people. In today's science and engineering research, teamwork and collaboration have become commonplace.

Enough has been said about effective team work, especially in business and management sections of bookstores online and in airports. For scientific research, however, collaboration often takes some unique conditions. We can summarize these conditions as: common goals, communication, complimentary skill set, and social cohesiveness.

A most common team setting is a student-advisor combination. There can be variations where an advisor is replaced by a committee member, and the student by a student group. In this setting, the student is often the main risk taker, taking the initiative in seeking out new research topics, conducting surveys, inventing algorithms, and carrying out the experiments needed if the work involved is empirical in nature. In this case, the advisor acts as a consultant, giving feedback to students from time to time and making sure that the student is on the right track. At the beginning, the advisor is often the generator of the initial ideas, defining an area where the advisor has experience in, and listing out several potential broad topics for the student to choose from. Later, the student becomes the main idea generator. In weekly meetings, for example, the advisor often asks "have you checked this paper out?" "Why not consider this other approach?" Or, "have you talked to so and so in our department, for he or she might have a solution for your problem?" When it is time for paper writing, the advisor often acts as an editor, especially for the first versions of the research paper. In fact, an effective advisor can help a student choose a better title, formulate a better abstract, and a better introduction, which are often the most difficult tasks to do for a beginner. See Section 5.6 for some other effective approaches with which your supervisor can help you improve your paper writing quickly.

These points are particularly important when dealing with multi-disciplinary research, where team members come from different disciplines. This is when your teammates are most likely your peers. Bear in mind that multi-disciplinary research is often the most rewarding, because it allows each researcher to "think outside the box" of his or her own research field. This can happen in bioinformatics, computational physics, or environmental sustainability fields. The authors had the pleasant experience of traveling far into China's west together with some of Asia's best ecologists and computer scientists in seeking ways to understand birds' migration patterns via GPS[5] trajectories. In a series of eye-and-ear opening workshops, researchers from different disciplines pleasantly discovered what each could do to achieve the goal of environmental sustainability.

In multi-disciplinary research, many new challenges exist. While it is fun to work with researchers elsewhere, it does take time to learn each other's objectives, tools, and terminologies. Sometimes it feels like getting a second Ph.D. degree. This period of re-education might last one or more years, after which one can find many exciting new horizons. Thus, we highly encourage Ph.D. students to be willing to drop in each other's seminars and conferences to learn something entirely new. You will be pleasantly surprised!

*   *   *   *

---

[5] Global Positioning System.

CHAPTER 5

Writing and Publishing Papers

In Chapter 3, you learned how to find novel and interesting research ideas and topics, and in Chapter 4, you learned how to conduct solid research to demonstrate your ideas. Assume that you have accumulated enough novel and significant results in your research. The next important task, for researchers and graduate students, is to write and publish high quality papers in order to tell the world about your contributions in research. Albert Einstein, for example, was awarded the 1921 Nobel Prize in Physics largely due to his published paper on the photoelectric effect (which gave rise to quantum theory) in 1905. In fact, Einstein published three other papers on Brownian motion, the special theory of relativity, and $E = mc^2$ in 1905. It is those four publications that contributed substantially to the foundation of modern physics and changed the world since.

In general, a research work and outcome can be very roughly categorized into top-notch (say top 5%), very good (top 5–25%), good (top 25–50%), and so-so (bottom 50%), based on its novelty and significance. Your effort (finding novel ideas and conducting solid research, as discussed in Chapters 3 and 4), is to "upgrade" your research outcome from so-so to good, to very good, and to top-notch, whenever possible. Which category a finished research shall fall into is, in fact, a highly complex issue. Experienced researchers, such as your supervisors, may have a better sense where your research outcome fits in. But let us assume that this can be done for now. Ideally, you should obtain at least good or very good research outcome before considering writing and publishing a paper.

We can also roughly categorize journals and conferences into the same four classes based on many factors (such as the Impact Factor, competitiveness, and so on). Again, this is a highly complex issue, but assume that this can be done. Then, good writing is to ensure that a research outcome will be published in a publication venue of the corresponding category. For example, a very good research outcome should be published in a very good publication venue, instead of being "downgraded" to good or even so-so ones, after repeatedly rejected by very good venues due to poor writing. Sadly, we have seen this type of down-grading happening too often in young researchers. Chapters 5 and 6 of this book will hopefully guide you to change this.

Most graduate students will have written essays and reports by the time they enter graduate programs. Some may have written newsletter articles, novels, poems, promotional flyers, and speeches. They will surely have written tons of emails, instant messages, and online chat. But most may have never written academic research papers, which are *quite different* from the other forms of writing mentioned above. As the late Turing Award winner and Nobel Lauriate, Herb Simon[1] was known to tell his students: to write a good paper, "you decide on what you want to say, say it, and then choose a good title!" There are many details embedded in this statement, which is the topic of this chapter. In particular, this chapter discusses the key elements of writing research papers, and emphasizes the main differences between scientific writing versus other forms of writing.

5.1   "PUBLISH OR PERISH"

As we have seen, publishing high-impact, top-quality papers in top, peer-reviewed venues in a sustained manner is extremely important for researchers, especially faculty members in universities and researchers in research labs. It is the most important means to share with the research community your contributions in research, and your papers will hopefully bring impact to the field. In addition, your academic promotion is also critically hinged on your high-impact publications. The same can be said about your research grants and projects. Thus, in the research circle, there is a phrase: "publish or perish"!

---

[1]Herbert A. Simon (1916–2001) was a professor at Carnegie Mellon University. He was a founder of many disciplines, including artificial intelligence. He received ACM's Turing Award in 1975 and the Nobel Memorial Prize in Economics in 1978.

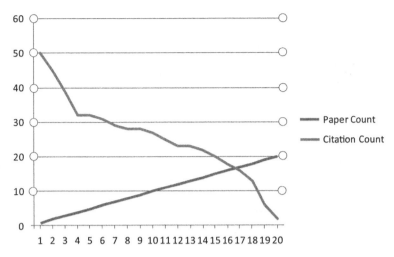

FIGURE 5.1: Illustrating h-index: an author's h-index is when the publication count line crosses the citation curve, given that the papers are sorted from more to less on the X-axis.

Indeed, the pressure for publishing high-impact papers is very high in universities in North America, and in many countries around the world. This pressure tends to have some adverse effects, including producing "salami" papers (see Chapter 3) that are not highly impactful and overlooking the other important responsibilities (such as teaching) of university professors. We are certainly against publishing many unimportant papers just to pad your resume.

Despite the negative connotations associated with the phrase "publish or perish," publishing high-quality papers that report novel and significant research results *is* the duty of researchers. Here, we emphasize that a researcher's goal is to publish high-impact papers that others can use to build their own research programs. One indication of impact is citation numbers, which counts the number of references others have made in their papers when referring to your work. Today we have many ways to count citations. Some count citations by journal articles only; for example, the Science Citation Index, or SCI, is an often-used index on journals from the Institute for Scientific Information (ISI), which is owned by Thomson Reuters. Instead of just counting a limited number of journals, other citation services may include most open-source publication venues, including conferences. Examples include the Google Scholar and the CiteSeerX services, which are increasingly being used to judge the impact of papers and researchers.

A particularly important measure is the h-index, which is a measure designed by Jorge E. Hirsch on both the productivity and the citation numbers of a paper. According to Wikipedia[2]: a scholar with an index of h has published h papers each of which has been cited in other papers at least h times. For example, a researcher with an h-index of 20 means that the researcher has 20 papers that are cited by others for 20 times or more, and the rest of the papers published by the researcher are cited less than 20 times. The h-index, as shown in Figure 5.1, reflects both the quantity of publications and the quality of publications, which can be viewed via a chart. We sort the publications of an author from the most cited to the least cited, and take this list as the X-axis. The Y-axis is the count value: we can plot the number of citations for each paper, which gives one curve (marked as "Citation Count"). Similarly, we can plot the diagonal straight line as the number of papers, which is denoted by "Paper Count." When the line and the curve cross each other, the corresponding X-value gives the h-index value (which is 17 in this example).

However, publishing top-quality papers is not an easy task. This leads to our next question: why is publishing top-quality papers hard?

5.2   WHY PUBLISHING TOP-QUALITY PAPERS IS HARD

There are usually two major avenues of publications: conferences and journals. Most good conferences and journals want to maintain their reputation of only publishing a limited number of top-quality papers. At the same time, most researchers want to publish in those top conferences and journals, because their papers will have a high visibility. Thus, the competition is very stiff.

[2] http://en.wikipedia.org/wiki/H-index

In computer science and some other disciplines, there are many top-rated conferences that publish full-length papers (e.g., 6 to 12 pages of double-column, condensed texts). The acceptance rate can range from 10 to 30%. Many top specialty journals in science and engineering also have an acceptance rate of about 10–30%. Some broad top-of-the-heap journals in science (such as *Science* and *Nature*) have very low acceptance rates.

Some students tend to "try their luck" when they submit their papers to those top conferences and journals. "30% is not too bad, and is much higher than winning a lottery!" They further explain mathematically: if I submit a paper to 20 conferences and journals with an acceptance rate of 30%, the probability that none of them accepts my paper is 0.7^{20}, which is about 0.014. Thus, the chance that at least one conference/journal accepts my paper is 98.6%!

We usually give those students the following advice: first, do not submit "lousy" papers to a good conference or journal just to try your luck, or just to get reviewer comments about the paper. Doing so will easily damage your reputation. Second, it is academic misconduct to simultaneously submit the same or very similar papers to multiple places (see Section 5.4 on plagiarism).

5.3 WHAT MAKES A GREAT PAPER?

What makes a great paper? What ingredients will ensure that your paper will get accepted? As you have probably guessed, there is no absolute answer to these questions; it is relative to the conference or journal you submit to, and the answer is also subjective by individual reviewers. When reviewers evaluate papers, they take into account the overall quality of the conferences and journals that you submit to. Despite these factors, there are some common elements that reviewers often use to judge and rank papers. We will review them below.

First, you need to understand the review process. Almost all conferences and journals recruit a pool of established researchers to review submitted papers. For each submission, 2–5 reviewers (usually 3) in the same or closely related areas are selected to review your paper. Sections 5.8 and 5.9 will discuss paper reviews for conferences and journals in detail. These reviewers often need to provide a set of "scores" for rating the quality of the paper, along with a detailed justification of their scores. As mentioned earlier, reviewers are always anonymous, thus, they often review papers with a critical eye. Scores are usually integers between 1 to 5, or between 1 to 10. The following is a typical list of questions that reviewers must answer with scores:

Please use an integer from 1 to 5 to answer each question below (1 for definitely NO, 5 for definitely YES):

1. Are the research and results novel?
2. Are the results significant?
3. Is the paper technically sound?
4. Is the paper clearly written and well presented?
5. Should the paper be accepted?
6. Are you highly confident about you.r review?

Clearly, Questions 1 and 2 are about novelty and significance of the research we have been talking about throughout this book. Question 3 is about research methodologies (see Chapter 4). The research must be technically sound to support the main claim of the paper. If the idea is not novel, the result is not significant, or the research method is flawed, the scores on Questions 1–3 would be low, and the chance for your paper to be accepted would be low, no matter how well you write your paper.

However, too often we see young researchers, including Ph.D. candidates, who have conducted novel and significant research, but have difficulty in writing their papers well. That is, their papers would receive a low score for Question 4 above. This also means that it is difficult and frustrating for reviewers to make an accurate judgment about Questions 1–3. For a high-quality and competitive conference or journal, reviewers are usually critical and cautious—if they are not sure about the novelty and significance of a paper, they tend to give low scores for Questions 1–3. Such a paper would then likely be rejected.

As we have been organizers of many competitive international conferences, often as Program Committee Chairs, we can say that among rejected submissions, *at least half of them were negatively influenced by poor writing.* This happens to many papers written by native English speakers. Chapter 6 describes many language-independent common misconceptions and flaws that often weaken the submitted papers significantly. If authors adhere to guidelines and methods that we propose in this and the next chapters, those papers would be better judged, and would have a better chance of being accepted.

We will start with several basic, untold truths about research papers.

5.4 A FEW UNTOLD TRUTHS ABOUT RESEARCH PAPERS

We would like to first lay down a few basic "ground truths" or postulates about research papers. These postulates are so basic and well accepted in the research community that your supervisors (or other researchers) may never explicitly spell them out for you. However, many early researchers learn these postulates in the hard way.

The first and most important ground truth about a research paper is that it must be truthful, honest, and accurate. Unlike commercial ads and product promotion, authors of research papers should be as unbiased as possible, discussing both strengths and weaknesses of their work in the paper. The experimental results must be truthful and accurate. Material, data, and results should be kept for a certain period of time (usually several years) after the paper is published for possible future verification by other researchers. If researchers intentionally falsify the results or mislead the readers, serious questions may be raised about the researchers' credibility; in the worst case, academic fraud may be charged. Occasionally we hear news about research misconduct where data and research results are fabricated, or the published results cannot be replicated. This often results in published papers being retracted from the journals, and/or researchers being dismissed from their research institutes. Honest mistakes are allowed, but strong due diligence should be applied to avoid mistakes that could have been avoided.

Another common form of research misconduct is plagiarism. This includes many forms. One of them is called citation plagiarism. This is when authors fail to cite and credit previous work that is similar to their work, and give readers an impression that the work is original and done by themselves. We understand that many authors may not wish to do this on purpose; often they may not find certain previous publications to know that their work has actually been done and published before. In many cases, reviewers can understand such unintentional omission, and will point out such previous papers. As authors, you must then study and cite these and other published papers, and discuss how your paper is different from them. If the previous work is indeed very similar or identical to yours, then you should not submit your paper anymore, until you improve it to show that it is better than the previous work, at least in some aspects.

In Section 3.2, we mentioned that you should do a thorough literature search, as thorough as you can before you start your research on a chosen topic or problem. This would help prevent you from wasting time and effort in "reinventing the wheel." When you make a claim in your paper on novelty, you may also say "to the best of our knowledge, our work is original . . ." to reflect your best effort.

Another form of plagiarism is multiple publications of the same or very similar content in different conferences and journals. Many conferences and journals explicitly ask authors to submit only original work, and forbid simultaneous submissions even if the authors intend to withdraw others should one submission be accepted. Multiple submissions and publications can be discovered easily (especially with modern electronic submission systems), and will easily damage authors' reputations.

When you include others' writings as your own, you run the risk of violating copyright law. Some graduate students copy many sentences or even several paragraphs from other papers, often from previous work or a general introduction to a topic, in their own papers. True, the description of previous work or a well-established topic is pretty much the same. But still, you cannot copy sentences from other papers into yours as if you wrote them. The "other papers" may actually include your own published papers. The publication agreement of your previous papers may stipulate that anyone, including you, needs to get a written permission if any part of the paper is reproduced. If you need to quote the whole sentences or paragraph as fair use in your paper, put them in quotation marks, and cite the sources.

An easy way to avoid copyright infringement, when you write similar content as in other papers, is to wait for a few hours (or days) after you read the other papers, and write the description in your own words. This way, your content will be similar, but the wording will be quite different.

The last often-untold truth about research papers is that it must reveal the full details of your research so that other researchers can replicate your results independently after reading your published paper. In Chapter 1, we described research as making novel and significant contributions with repeatable and verifiable results. Thus, it is your responsibility to reveal technical details of your work as fully and as clearly as possible in your paper. If other researchers require additional details of your published paper, you are obliged to supply them. Indeed, research in science is "self-regulating" and "self-correcting," where published results and papers will be analyzed, replicated, verified, and improved upon by other researchers. It is an open (fully revealing), fair (relatively speaking), and free (i.e., academic freedom) community. If your work contains trade secrets or methods with potential commercial gain, you can choose not to publish it, or you can secure patents or other intellectual property protections before the paper is submitted.

When you are ready to write your first papers, you should certainly get your supervisor involved. You may have a proofreader (e.g., a native speaker) to review and revise the paper. What are the roles of you (Ph.D. candidate), your supervisor, and the proofreader in a paper writing process?

5.5   THE ROLES OF YOU, YOUR SUPERVISOR, AND PROOFREADER

Quite a few graduate students are from abroad, where their native language is not English (such as Chinese, Persian, Spanish), but they need to write papers in English. "Since my native language is not English, my supervisor should write the papers, while I am doing the experiments," some may say. Or, students just write a first draft, give it to their supervisor or proofreader to revise and polish the paper to "perfection." These viewpoints are certainly not correct. You must learn to write papers well by yourself by the end of your Ph.D. study, regardless of whether your native language is English or not.

In fact, many native English speakers do not know how to write research papers well. One of our Master's students, who is a native English-speaking Canadian, almost failed the thesis defense due to his poor writing. There are many language-independent misconceptions and flaws (see Chapter 6) in writing research papers, and they are often harder to overcome than grammatical errors. Your supervisor will play an important role in correcting them.

A good understanding of the roles of you, your supervisor, and proofreader is important in writing a high-quality research paper. Here is a list of roles that the supervisor may play in paper writing:

- Help you to understand the goals, logic flow, organization, and arguments in writing research papers.
- Help you to identify specific misconceptions in writing research papers, and you must learn to correct them quickly.
- Make high-level suggestions on how to improve the presentation and argument of the papers.
- Make a final check of the paper before it is submitted.

Your role in paper writing:

- Work with your supervisor closely for the first 1–2 papers, and learn to write papers well quickly (see Section 5.6 for advice for supervisors).
- Become independent quickly, and be able to write the whole paper well by yourself.

Proofreader's role:

- After a paper is largely finished (with excellent structure, flow, argument, organization, and presentation) by you and your supervisor, a proofreader can be asked to further improve the paper and correct minor English errors. Other graduate students in the group or in the department would be good candidates.

We want to emphasize that the role of proofreaders, unless they are also researchers in the similar area, is quite minimal. Often they cannot help you strengthen the logical flow, arguments, or organization of your paper. They can only point out minor English errors (such as the question of whether "the" should or should not be used in a certain place, and the use of prepositions). You and your supervisor are most crucial for writing a high-quality research paper.

So, how can you and your supervisor can work together most effectively in improving your paper writing? We answer this question in the next section.

5.6 SUPERVISORS: HOW TO IMPROVE YOUR STUDENTS' WRITING MOST EFFECTIVELY

This section is mainly written for supervisors, who wish to improve their graduate students' writing most effectively. If you are a graduate student and want to work with your supervisors to improve your writing skills quickly, talk to your supervisors to see if they can use the method we suggest here.

Many supervisors first ask their students to write paper drafts, and correct them on papers (e.g., with red ink). They pass the corrected drafts to students to "input" the corrections in the paper. This process may be repeated many times. Unbeknownst to supervisors, their students often have a hard time reading their handwriting, and do not know *why these changes have been made*. The improvement in the students' writing skill is often minimal in each such writing-revision iteration.

We have found that the following method can very quickly improve students' writing skill, to the point that they can write well by themselves after writing only 1 to 2 papers. We call the method the PI method, or the Progressive Improvement method. It may take a bit more time in the beginning, but the extra effort at the beginning will be paid back many times during their Ph.D. study.

The PI method works roughly as follows. When it is time to write a research paper, you can ask your student to write a few sections first, or about half of the first draft. These sections may include the Abstract, Introduction, and a section on your new theory or methods. Why not ask him to write the whole paper? Well, if this is the student's first paper, it is likely that the draft needs be completely rewritten. Thus, the student does not need to waste too much time in writing a complete draft that will not be used in the future.

You then sit with your student, and work closely with him, *word by word*, for the first page (just one page) of the paper. For a student who is a native English speaker, you may focus on high-level structure, logic flow, and how to make convincing arguments in the paper. For a foreign student, you may also need to work on detailed sentence structures and choice of words. You need to provide detailed explanations about why you revise the paper this way. Ask your student to write down each writing principle and each type of mistake your student made. You then tell your student to *learn and generalize* from these principles and mistakes, and apply them to revising the rest of the paper (and to write new sections). You ask him to review the notes often, take time to revise the paper carefully, and try not to make the same mistakes again.

Your student may come back in a few days with the revised paper. You can first quickly review the previous page(s) you have worked on, and offer further advice on improvements. Then you work with him, word by word, on the next new page (still, just one page). You should notice that the number of mistakes and needed improvements might be cut by half. If your student makes the same mistake as the last time, you should give him a serious warning. Your student should write down new tips and mistakes, and then go away to revise the rest of the paper, and come back to you in a few days. This process goes on progressively (thus called Progressive Improvement), until the whole paper is improved once. By that time, the writing skill of your student should have improved tremendously.

It is often necessary to go over this whole process two or more times. In the second and third rounds you and your student may focus more on the overall structure and logic flow of the paper, adding experiments to make the results more convincing, adding examples to make it easier to understand, and so on. Again, you work closely with your student, but only on 1–2 issues, or only on 1–2 pages at a time. Then, you may ask him to apply the ideas to improve the whole paper. Several such iterations are often needed to produce a highly polished and satisfactory paper, to be submitted to competitive conferences or journals. By the time the first paper is finished (usually in several months), you student should improve his writing tremendously, and be ready to write a reasonably good paper by himself.

Your student will also learn how you have tried to make their paper as perfect as possible. Keep each version of the revisions so that your student can see how many changes have been made, and learn from this process after the paper is submitted.

In Chinese there is a saying: it is better to teach people to fish, rather than giving them a fish. (The conventional English translation of this is: Give a man a fish and you feed him for a day. Teach a man to fish and you feed him for a lifetime.) By spending some extra time teaching your students to write their first paper at the beginning of their graduate study, they learn to write good papers by themselves for the rest of their research career and life. This in turn frees your time for working on other important research problems.

Often, we have several graduate students who are learning to write papers, and we use the following methods to be even more efficient. We will distribute the first draft of a paper to these students, and ask them to review and revise the first page carefully on their own. We will then use Skype (a software for online group voice chat) and Teamviewer (a software that allows several people to view the same screen where the paper is actually revised) to discuss and revise the paper simultaneously. This way, all of our graduate students can contribute, revise, and learn how to write better research papers, even when we are in different places or countries.

Below, we will first discuss where to submit the paper after it is written.

5.7   WHERE TO SUBMIT: CONFERENCE OR JOURNALS?

In some disciplines, such as biology and statistics, conference papers are usually very short and in the form of an extended abstract, and the acceptance rates are usually high. The purpose of the conferences in these disciplines is to let research-ers meet and present their most recent work, usually as posters. Often thousands of researchers go to such conferences each year. Complete work is often published only in journals. In some other disciplines, such as Computer Science, confer-ence papers are full-length with 6 to 12 pages of condensed text, often in double columns, reporting on novel contributions. The reviews for such conferences are more rigorous, and thus, the acceptance rates are typically quite low. If a paper is accepted, authors often have 20 to 30 minutes to present their work orally in a conference session, sometimes as a poster as well. Often, such conference papers are regarded as important as journal paper publications, although different depart-ments, schools, and universities may put them under different weights.

Here are some other differences between full-length conference and journal papers:

- Conference papers often have a shorter review time than journals, and their review processes are also different (see Sections 5.8 and 5.9). Thus, conference papers are suitable for reporting the most current work with rapid progress. Going to conferences also allows you to meet other researchers and build up your academic network.

- Annual conferences have submission deadlines while journals do not. Thus, conference deadlines are often taken as *drivers and landmarks* for making progress in research.

- Conference papers, including the full-length ones, usually have page limits, while most journals do not. Thus, journal papers are usually longer, extending full-length conference papers to report complete work, and thus are often regarded as "archival."

One way to enter a research field is to attend a top or main conference in that field. In computer science, for example, each subfield such as artificial intel-ligence or machine learning has its own "top conference." Here you will meet with the leaders of the field, hear the latest progress in the field, and meet numerous people who are active researchers in the field. Soon, you will want to be an author yourself, and start to look for a conference to submit your paper to. If your paper is accepted, you can present your paper in conferences while others listen and learn about your work. It will be a boost to your confidence in research, and make you a mature researcher.

Which conferences and journals will you select to submit your papers? Each discipline has a set of conferences and journals; usually some are considered top-tiered, while others so-so. You have probably heard about some humorous (yet true) stories about nonsensical computer-generated papers being accepted by peer-reviewed conferences and journals. You certainly do not want to submit your papers there!

Here are some general guidelines in choosing conferences and journals to publish your work. If a conference is the main conference in the field where most important researchers go, where "importance" can be measured by citation numbers, h-index, or leadership in an area, then you may consider submiting a paper or poster even if the acceptance rate is high. Otherwise, choose the main conferences in a field with an acceptance rate around or below 1/3. For journals, it is hard to specify an Impact Factor (IF) as a guide to determine where to submit your papers, as IF varies a lot for different disciplines. One general guideline is to choose the top tier, or the top 25%, of the journals in the field in terms of IF. You may also talk to your supervisor who should know well which conferences and journals belong to top-tier.

After you submit a full-length paper to the conference, you might ask: how are conference papers reviewed?

5.8 HOW ARE FULL-LENGTH CONFERENCE PAPERS REVIEWED?

Many prestigious and competitive conferences in computer science require that the names and affiliations of authors remain anonymous in the submission, such that the reviews can be more fair—reviewers will judge the papers only based on the contents, not the authors. Some conferences, however, do not have this requirement precisely because author information can also provide some creditability of the work. In both cases, it is usually quite fair when your submissions are reviewed by other researchers.

Let us consider the process in which a top conference reviews and accepts full-length papers. The authors of this book have been conference reviewers, program committee (PC) members, senior PC members, PC chairs and conference chairs. Naturally, we are in a good position to answer this question.

First, when the conference paper deadline closes, typically the conference chairs will start to filter out the papers that obviously do not make the cut. Some are rejected right away due to their wrong format, such as being longer than the page limit. Typically, there are also a number of senior PC (SPC) members; each oversees the review process for a subset of the papers and provides comments and recommendations for PC chairs at the end. A paper is often reviewed by three reviewers, who are at arm's length from the authors. Often, this means that the reviewers are not current and former colleagues, students, advisors and friends of the author.

What do reviewers look for when they review a paper? Section 5.3 described briefly what makes a great paper. Again, the most important attributes are novelty and significance, as mentioned many times in the book. Reviewers would want to find out what is new in this paper, compared to established literature and practice in the field. If they cannot find what is new about the paper, say in the abstract and introduction of your paper, they might start looking for reasons to reject the paper. Another important thing in paper review is to look for evidence to back up what the author claims are the innovative aspects of the paper, to ensure that there is real impact and significance in the research. For example, the author might claim that the paper presents a new method for learning communities in a large-scale social network in the abstract and introduction of the paper. However, if the experiments deliver only tests on a small-scale data set, or if the experiments do not compare with some well-known methods in the literature, then there will be enough grounds to reject the paper. Section 5.3 discussed what makes a great paper, and reviewers need to give scores to assess a paper's originality, significance, technicality, and presentation.

Generally, to reject a paper is often easier than to accept one, because there are many reasons a paper can be rejected, and only one of them needs to be found. In other words, a paper might have many critical points where things can go wrong, and if one of these critical points is not handled properly, the paper might be rejected.

We can simulate the process of rejecting a submitted paper by pretending that we are reading a freshly received submission. First, a reviewer can read the abstract of the paper and search for the term "In this paper . . ." or an anchor place where the authors state what their main contributions are. The reviewer can then take this contribution statement and start looking for elaborations (that is, by following the top-down refinement discussed in Chapter 6), beginning with a reading of the introduction section. If the reviewer cannot find any elaboration, it is a sign that the paper was not written in a logical way, and it is very likely that the paper is poorly written. If similar patterns can be found in many other parts of the paper that shows that the paper contains too many logical confusions or syntactic mistakes, then the reviewer may consider these facts as strong evidence that the paper should be rejected for a thorough rewrite. Subsequently, if the paper clearly states its innovations in both the abstract and introduction sections of the paper, but if the author has not followed up with strong evidence to back up these contribution claims (i.e., the author did not state how he did his solid research, as we mentioned in Chapter 4), then the paper again can be rejected. If the research results can be in place, but the author has not discussed or cited some very relevant works in related literature, then it may be a sign that the author is not sufficiently familiar with the established literature in the field, and it may be another ground to consider this as a major weakness, which may lead to a rejection.

Summing up the above procedure, the process of a quick scan on a paper can be rather strict, but fast: it may take 20 minutes for a reviewer to check all of the above points, before a rejection decision is made due to a lack of logical flow, clarity, sufficient related works or research results. If, however, the paper survives the above screening process, it may still not mean that the paper will be accepted. The reviewer might take another 30 minutes or longer to carefully read the technical parts of the paper, go through the derivations, theorems, and results, check the relevant literature, and perhaps even ask the opinion of another expert in the area, before giving a recommendation on the paper.

In recent years a reviewer's workload is increasing each year. For example, a typical computer science conference can easily receive nearly 1,000 submissions. A reviewer might have to review 10 papers in several weeks. The time they spend on each paper is usually rather short. Each reviewer needs to provide the reviewing scores, as well as detailed justification and suggestions. In Chapter 6 we will discuss how to make papers easy to read, so that your paper can pass the so-called "10/30" test.

Some conferences allow authors to view the reviews and input their feedback online before final decisions are made. If such an opportunity is given, you should try your best to provide succinct, clear, and strong statements about reviews, especially if you believe some aspects of the reviews are not correct.

The final acceptance decision of each paper will be made by the senior program committee member (SPC) and the program chairs. Once the decision is made in a conference, which is usually just "Accept" or "Reject," the decision is final, and usually cannot be appealed. There is no room to formally reply to the reviews and to submit a revised version, as most journals would allow. So, how are journal papers reviewed?

5.9   HOW ARE JOURNAL PAPERS REVIEWED?

Journal papers are peer-reviewed by other researchers, in a similar way as a conference paper. The EIC (Editors-in-Chief) of the journal may assign each submission to one of the Associate Editors (AE) of the journal. The AE in charge of your paper will recruit reviewers and oversee the reviews of your submitted paper. The final decision is usually made by the EIC.

Compared to conference papers, major differences exist that make journal papers much "deeper." First, a journal usually does not have a very strict page limit, making it possible for authors to fully express themselves. Thus, authors can go much deeper in a journal paper than in a conference paper in describing their research results. Second, the process of paper review for a journal paper is often longer than a conference paper, going through several "rounds" of comments and response between the reviewers and authors. This process makes it possible for authors to submit a revised paper to address all reviewers' concerns.

More specifically, after the first round of review, authors may receive one of the following four decisions: accept as is, accept with minor revisions, accept with major revisions, and reject. In the middle two cases, authors need to prepare a revision within a certain time limit to resubmit to the journal. Accompanying the revised manuscript, authors also need to prepare *a clear and detailed point-to-point reply to all reviewers' comments*, especially on the negative ones, in a response letter. Here authors can give a detailed rebuttal if they believe that the reviewers are wrong. When authors believe that the reviewers have made good suggestions, they should improve the paper, and specify how and where in the article they have addressed these comments. The response letter can be long (e.g., 10 pages), and should be carefully written.

If the paper is initially accepted with major revision, after the revised paper is submitted, the paper will usually be handled by the same AE (Associate Editor), and reviewed by the same reviewers. The second-round revision can still be rejected, if reviewers are not convinced and satisfied with your revision, or it might result in a second major revision (only in rare cases), minor revision, or acceptance (happy ending!). If the second-round decision is a minor revision, you should submit the revised paper, and another point-to-point response letter, similar to the one for the major revision; but this one can be shorter than that for a major revision. Usually, the revised paper will be checked by the associate editor in charge, along with a few designated reviewers, only to see if all of the minor concerns are addressed. If so, the paper will be accepted.

Occasionally, during these rounds of review you may think that you are being treated unfairly. In that case, you can always write a letter to the EIC to complain.

Compared to conference papers, the process of submitting and revising a journal article is often lengthy, but it provides very good training for a student and researcher alike. It is in this process that more detailed weaknesses of the manuscript can be brought out and addressed. Thus, despite the lengthier time delay in getting a paper published, it is still worthwhile to submit an extended version of a conference article to a journal, in order to "close the book" after fully exploring an idea. For this reason, journals are often considered to be archival.

When extending a full-length conference paper to a journal article, how much extra material is often included? On this question, the answer varies from journal to journal. In some conferences, such as the computer graphics conference ACM SIGGraph,[3] most conference articles can be directly transferred to a corresponding journal, such as the ACM Transactions on Graphics journal in this example. This is done under the agreement between the conference and journal on copyright matters, and on how papers are cited, so that one of the conference or journal articles will be cited at a time and citations will not be counted twice. In other cases, full-length conference papers need to add at least an additional 25 to 30% of new material before being considered for journal publications. This additional material can be more experimental results, more theoretical results, more discussions and comparisons with related works, or a combination of the above. In this case, we suggest that you write a cover letter to the EIC detailing what is new in the journal submission compared to the published conference paper.

As you can see, the process of paper review in a typical journal is quite different from that in conference. While review of conference papers is a "one-shot" process, meaning that the review results cannot be disputed once the final decision is made, for a journal paper the authors can often engage in rounds of communi-

---

[3] Association of Computing Machinery, Special Interest Group on Graphics and Interactive Techniques; http://www.siggraph.org

cations via the response letters, and thus the decision process is more interactive. In addition, because a journal is often longer in page count, authors can write more thoroughly about a subject. Thus, we strongly suggest that Ph.D. candidates should try their best to submit and publish 1–2 top-quality journal papers reflecting their Ph.D. work, especially if they plan to enter the academia.

CHAPTER 6

Misconceptions and Tips for Paper Writing

In the previous chapter, we have given an overview on the process of paper writing, submission, and reviewing. In this chapter, we focus on some common and major misconceptions and flaws that junior researchers and graduate students often have when they write their first research papers. These flaws are *language-independent*—that is, native English speakers often make them too. Ironically, it is often much harder to correct those language-independent flaws than grammatical errors in English, and thus, supervisors may need to be forceful in correcting them, and correcting them early, before the Ph.D. thesis is written. Publishing a series of papers in competitive journals and conferences during Ph.D. study is the best way to validate not only the thesis topic, but also the good presentation style of the papers.

6.1 "IT'S SO OBVIOUS THAT OUR PAPER IS GREAT"

One common weakness of papers is that the authors do not make a strong argument for the research in their paper. They might think that the reviewers can see and conclude easily that the paper is great. This may be due to the culture and environment that they grew up in; e.g., some cultures believe that "we should be modest," and thus we should not stress the importance of our work too much. However, as we saw earlier, publishing in top journals and conferences is highly competitive, and thus it is your responsibility and to your benefit to argue, *as strongly as you can but not exaggerate*, that your work is novel and significant.

Indeed, *a research paper is essentially an argument about one central theme*: our research work makes novel and significant contributions. This central theme is almost always supported by the following logical steps:

- **The problem is important in advancing knowledge.** If the problem you study is not important, then why study and write about it?

- **Previous works A, B, . . . have been done, but they have certain weaknesses.** Most likely some previous works have been done, and if they are already perfect, there is no need for you to study it.

- **We propose a new theory/method/design/process Z.** You should emphasize the novelty of your work. Is it the first time that Z is proposed? Is Z intriguing and surprising? If so, you should say it in the paper.

- **We prove/demonstrate that Z is better (at least in some aspects) compared to A, B, and others.** Can you prove it theoretically? Do you run extensive experiments to demonstrate the superiority of Z compared to previous works and the state-of-the-art (see Chapter 4)? This supports the high impact and significance of your work.

- **Discuss Z's strengths and weaknesses.** You should also be upfront about the weaknesses of Z. A research paper is not like product promotion—it must be honest, fair, and accurate. The weakness also leads to future work of Z, usually described in the Conclusion section of the paper.

Each component should be further supported and justified, as strongly as you can, by your paper. For example, when you claim that the problem is important, you can support it by showing who else also claimed this importance in the past, citing many previous publications on the problem, or referring to its use in real-world designs, engineering work, and applications. When you claim that your new theory or method is better than a previous one, you should be able to prove it theoretically, or run extensive experimental comparisons and perform statistical significance tests on the results (see Chapter 4 for how to justify the superiority of your work). If you show that your new method has been deployed in practice, such as industrial applications, it would be even better. The more support you have for your argument, the more convincing the paper will be, and the greater the chance that the paper would be accepted.

If you are the first one to propose the new theory, paradigm, or method, you should claim it in your paper: "To the best of our knowledge, we are the first to propose . . .". Be explicit and forceful, and avoid ambiguity. Use the active-voice sentence structure to describe what you have done (we propose . . . ; we demonstrate . . .); a passive voice (it was proposed . . .) is confusing as it is unclear who actually proposed it. Make sure to distinguish what your paper is proposing and solving from what previous works have proposed or solved.

Note that making a strong argument for your paper does not mean that you should exaggerate your claims. In fact, you should never claim anything more than what you can support and justify in the paper. Opposite from the statement "I am modest," we sometimes see the mistake of claiming "I am the greatest." Some authors make a grand claim (such as "We have solved the AI problems completely"), but grossly fail to support it. This casts a very negative opinion about the paper, which can be easily and quickly rejected.

Note that while making a strong argument about the superiority your research, use a polite tone, such as "to the best of our knowledge," "as far as we know." Similarly, use a polite tone when discussing the weaknesses of previous works, such as "it seems that the previous work . . .". Back up your claims with strong evidence.

If you plan to submit your paper to a specialized conference or journal, you may not have to say, literarily, that the problem we are studying in this paper is important. This can be implied in the description of the problems. But none of the above logical components should be missing in the paper.

To summarize, your paper is an argument, and should have one central theme: your paper contains novel and significant contributions. You should emphasize your contributions. You should argue for your paper, as strongly as you can, and at the same time, not exaggerate your contributions. It is a tricky balance. Be positive, accurate, upfront, and balanced.

6.2   "IT IS YOUR RESPONSIBILITY TO UNDERSTAND MY PAPER"

Very often, when early researchers (including Ph.D. candidates) write their papers, they unintentionally make their papers hard to understand by reviewers and future readers. They often believe that it is the reviewers' responsibility to understand their papers (hence the title of this section). They often blame reviewers for not having enough knowledge or putting in enough time to understand their papers ("Reviewers should read my paper more carefully and check out my published papers cited, to fully understand this paper." Or "Reviewers are so careless. They point out . . . but it is clearly stated at line 23 of the right column on page 5!"). Furthermore, some researchers even think it is a "good thing" to make their papers hard to read ("If my paper is easy to understand, it must be too simple!"). They might think that it is natural for research papers to be hard to understand ("My paper should be hard to understand—it is based on many years of research and my Ph.D. thesis!").

In the last section, we describe one objective of writing a paper: to make an argument, as strongly as you can, for the novelty and significance of your paper. Here, we want to add another objective: you *need to do so as clearly and as simply as you can*. The two objectives are, surprisingly, not contradictory to each other.

Why is the objective of clarity so important? A major reason is that reviewers are very busy researchers themselves, and they won't spend hours to understand every page of your paper. If reviewers don't quite understand your paper due to your poor writing, they tend to be frustrated, and give you low scores and negative recommendations (Section 5.3). The other reason is that after your paper is published, as we discussed in Section 3.3, readers, who themselves are researchers, need to get your main ideas and results quickly. Research papers should be written in a straightforward, direct, simple, and clear manner anyway. It is a significant milestone in a graduate student's career when he or she can consider "This paper is so simple," as a compliment rather than a criticism, as long as the simpler description addresses the same problem. As we will show you, it is actually not too hard to do.

6.3   THE 10/30 TEST

Whenever you are writing a research paper, or a report or thesis for that matter, you must remember at all times that *you are writing for your readers*, such as reviewers, other researchers, and thesis examiners. You know your work so well after many years of research, but this is not true for your reviewers. Often it is because you know your work so well that it is hard for you to write for others. You must put yourself in reviewers' shoes, so to speak. This is the same as designing products for your target clients and customers—you must think from the perspectives of your end users.

To write for reviewers, you must first gauge the knowledge level of average reviewers. If you are submitting your paper to a general conference or journal, a gentle introduction to the problem you study is needed. You should assume that the reviewers have little knowledge on the specific research problems you work on. You should certainly assume that reviewers have never attended your weekly research meetings, nor have they heard your presentations on the work! All they have at hand is this paper that they are reading. If they have questions while reviewing your paper, they cannot ask and communicate with you!

One way to check if your paper is clearly written is to judge if it can pass what we call the "10/30 test": for average reviewers, can they get a good idea of what the problem you are studying is and what your main contributions are within 10 minutes? In addition, can they get a very good understanding of your paper (assuming a 10-page conference paper) to make an acceptance decision within a further 30 minutes? Most likely reviewers will not spend more than one to two hours to review your paper (including writing the reviews).

Before you submit your paper, give it to some colleagues who are familiar with the area but have not worked on your problems, to see if it can pass the 10/30 test with them!

It is somewhat a "dilemma" that the more novel your work is (which is a good thing), the harder that you can convince the reviewers. You must introduce your work "gently" and support it strongly with technical details and convincing results (such as definitions, theorems, proof, designs, methods, data, experiments, and applications). How can you do these two "opposite" tasks well? The question turns out to be a key point in writing research papers, and we will present it in the next section.

6.4   TOP-DOWN REFINEMENT OF PAPERS

Top-down refinement, briefly speaking, means that you must present your work in the general-to-specific style. This can be facilitated by a top-down section structure of papers. Recall that, in Section 3.3, we discussed a typical structure of a research paper (duplicated here in Figure 6.1) and described how such a structure can help you to read the papers quickly to get the main ideas and results. You should write your paper in this way too, so that others, especially reviewers, can get the main ideas and results of your paper quickly and easily.

Let us now explain this key issue in more detail. First, recall that your paper should have the following logical steps supporting the central theme, briefly,

- The problem is important
- Previous works have been done but they have certain weaknesses
- We propose our new theory/method/design/process
- We prove/demonstrate that ours is better than previous works

In your research paper, you need to "tell your story" to reviewers of these central criteria several times (4 times, actually) with different levels of details. That is, you need to tell your story in 10 words (as in the title), 200 words (in Abstract), 1,000 words (in Introduction), and 5,000 words (the main body), with increasing levels of details. More specifically:

Title: The title of the paper is the highest level of summary of the central theme. It may contain a few to a dozen or so words (say 10 words), and it must be at a very high level (i.e., not too technical). It should convey a positive and exciting tone by using, for example, words such as "improving," "novel," or the "hot" topic that your paper is about, such as "social networks."

If writing a paper is hard work, then creating a good title is an art. The title of your paper gives the reader a first impression, and this is why some seasoned researchers even go as far as say that the title equals half the paper. A first rule of thumb is that you want your paper to stand out in a pool of many other papers of the same type, either as a paper in an issue of a journal, which might consist of 10 or 20 papers, or in conference proceedings with 100 or more papers. Another rule of thumb to follow is that your title is no longer than one line; a long title gives an indication that the work might be too specific, or the authors are not good at summarizing their work. Another rule of thumb is to ensure that no one has published the same or very similar title before. This rule can be enforced by checking on a search engine, which often returns a list of closely matched titles. You may try to enter some potential titles in a search box and see how many hits you will get as a result, and whether the returned results resemble the paper you are writing.

Abstract: The Abstract of the paper may contain a few hundreds of words (say 200 words). It must be *a high-level 200-word summary of the complete central theme*, telling readers about your work. In Figure 6.1, we use a "200-word story" and "elevator pitch" to describe the Abstract. That is, the Abstract must be at a high level, positive, simple, and it tells a complete "story" about your work in 200 words. It should attract people's attention, like an elevator pitch when you tell investors (reviewers here) how great your product (your paper) is within a few minutes.

Here is a sample Abstract. We add some comments in [. . .].

General web search engines, such as Google and Bing, play an important role in people's life. [The problem is important] However, most of them return a flat list of webpages based only on keywords search. [Previous works have certain weaknesses] It would be ideal if hierarchical browsing on topics and keyword

Title — A 10-word highest-level summary

Abstract — A 200 -word story (elevator pitch): very high level. Emphasize your contributions

1. Introduction
2. Previous Works — A 2 -page story: high level. Emphasize background & motivations
3. A New Method for ...
    3.1 Framework of ...
    3.2 Major Components of ... — Expand on various parts: **Top-down ref at all levels!**
        3.2.1 Feature Extraction
        3.2.2 SVM with a New Kernel Function
    3.3 Time and Space Complexity
4. Experiments
    4.1 Comparing with Previous Methods
    4.2 Parameter optimization
    4.3 Discussions — A 200 -word story: Summary & future work
5. Conclusions
    Appendix   Charles X. Ling — Detailed proof, etc.

FIGURE 6.1: Outline of a sample paper, and an illustration of top-down refinement.

search could be seamlessly combined. In this paper we report our attempt toward building an integrated web search engine with a topic hierarchy. We implement a hierarchical classification system and embed it in our search engine. [A very high-level summary of our work] We also design a novel user interface that allows users to dynamically request for more results when searching any category of the hierarchy. [Emphasizing novelty and usefulness] Though the coverage of our current search engine is still small, [be upfront on the weakness of the work] the results, including a user study, have shown that it is better than the flat search engine, and it has a great potential as the next-generation search engine. [Using positive words to show the significance and impact of the work.]

As you can see, the abstract presents a complete line of argument of the central theme at a very high level without using technical jargons. It also casts a positive and exciting tone about the novelty and significance the work.

Introduction: The Introduction section usually takes 1–4 pages (on average, 2 pages) to write. In the Introduction, you will *re-tell your central theme* again with more explanations on every component of the argument than the Abstract.

The Introduction should also emphasize the background, namely, why the problem is important; who has studied it; who should read the paper, and where have similar techniques been used, etc. The Introduction should still be written at a high level, avoiding technical terms and details as much as you can.

What is the relation between the arguments, such as "the problem is important," in the Abstract and Introduction sections? Simple! *Each argument in the Introduction is simply a refinement of a corresponding one in the Abstract, in the same order.* A simple and rough rule of thumb is that each sentence in the Abstract can be expanded and refined into roughly three to 20 sentences in the Introduction (20 sentences can form a paragraph). If the Abstract consists of sentences A, B, C, . . . , then the Introduction will be A1, A2, . . . B1, B2, . . . C1, C2, . . . The set of sentences Ai, i=1, 2, . . . , is simply an expansion of A. The same applies to B, C, and so on. If the sentences and logic flow in the Abstract are modified in a revision, then be sure to remember that subsequent revisions should be made in the Introduction to reflect the changes.

Here are the beginning sentences in the Introduction that accompanies the sample Abstract we mentioned earlier in this section. Again, our comments are included in [. . .].

General web search engines (such as Google and Bing) play an important role in people's lives in helping people to find information they want. According to SEMPO (2009), more than 14 billion searches are conducted globally in each month. However, most of them return a flat list of webpages based only on keywords search. As queries are usually related to topics, simple keyword queries are often insufficient to express such topics as keywords. Many keywords (such as china, chip) are also ambiguous without a topic. It would be ideal if hierarchical browsing and keyword search can be seamlessly combined, to narrow down the scope of search, and to remove ambiguity in keywords. The original Yahoo! Search Engine is a topic directory with hierarchy for easy browsing. However, . . . [background about search engine development]

You can see that the first two sentences in the Introduction are further elaborations of the first sentence in the Abstract. The following sentences are further elaborations of the second and third sentences in the Abstract section. The Introduction also provides the background on search engines, and the motivation and rationale of this research.

At the end of the Introduction you usually describe the layout of the rest of the paper. You tell the readers what each section of the paper will describe very briefly. Be sure to use an active voice as much as you can. For example: We organize the rest of the paper as follows. In Section 2, we discuss previous work on . . . In Section 3 we describe. . . . Finally, in Section 4 we show . . . etc.

Previous work: You may have a section on reviewing previous works, such as Section 2 in Figure 6.1. This is a further expansion of the corresponding part in the Introduction, which in turn expands from the Abstract. The key here is to demonstrate to the readers that you are familiar with the state of the art, so that your further statement on your contributions has credibility. Reviews of previous works do not need to be long, but you must draw the difference between previous works and your work.

Sections on your work: Your main original work will be described here. As it may be quite lengthy, you may need several sections for it. These sections make a detailed argument on your central theme. These sections describe what your novel theories, methods, or processes are (Section 3 in Figure 6.1) and why they are significant; that is, why they are better than previous works and the state-of-the-art (Section 4 in Figure 6. 1). You can have more sections (such as new theory, deployment, and so on) for your work. The overall structure of these sections is simply a further top-down refinement of the corresponding parts in the Introduction.

In fact, you should apply the top-down refinement *at all levels* in these sections. Your top-down refinement should be supported by the organization of the paper. That is, you create high-level subsections (such as Sections 3.1 and 3.2), to describe the high-level ideas, and then use low-level subsections (such as Sections 3.1.1, 3.1.2) to describe more technical details, as shown in Figure 6.1.

The top-down refinement is often used within each paragraph. Usually one paragraph should have one central idea. It often starts with a summary sentence (such as "Active learning has been shown to reduce the number of labeled examples significantly"). Then you further elaborate this summary sentence by showing why and how in the rest of the paragraph.

Conclusion: After you describe your contributions in great length, near the end of your paper, you need to write a summary or conclusion section. In this section, you re-tell the whole story again at a high level. The Conclusion section can be similar to the Abstract, with more details on your novel and significant contributions now that you have presented all the specifics, and future works.

With this top-down refinement structure, it becomes very easy for reviewers, and future readers, to understand your ideas. At the same time, they can go deeper into your technical details and results to any level of detail they wish. Reviewers can spend 10 minutes on the Abstract, Introduction, and some top-level descriptions to understand the main contributions of your work at a high level. This helps them understand quickly the rest of the paper by spending 30 minutes or so to hover between high-level subsections, and if they want to, go deeper into technical details at low-level subsections. Only this way can your paper possibly pass the "10/30 test" described earlier in the section.

Do you find that our book is easy to understand? That is because of how we framed it: we have been using the top-down refinement in writing this chapter, and the whole book.Now, please do the same in your papers!

6.5   CREATE A HIERARCHY OF SUBSECTIONS AND CHOOSE SECTION TITLES CAREFULLY

As we have seen the importance of writing research papers in a top-down refinement fashion, it is thus extremely important to create levels of (sub)sections, and choose section titles very carefully to reflect the overall logic structure of the paper. Note that in some fields and journals, papers often have a fixed structure and section titles (such as sections on Introduction, Hypothesis, Methods, Results, and so on). If this is the case, you should follow the convention. Avoid a very long section without subsections. For example, if the Review section is very long, split it into several subsections (Sections 2.1, 2.2, etc.), each perhaps discussing one type of previous work. On the other hand, avoid very deep subsections (such as 3.1.2.1.1). Usually you should not go deeper than 3 or 4 levels in structure. If you have to use deeper subsections, consider including two or more first-level sections (for example, Section 3 for the new theory and Section 4 for the new method) rather than putting them into one big section.

At the beginning of each section and subsection, you should write a brief summary or introduction to this section. Thus, between Section 3 and 3.1, there should be some introductory or summary paragraph(s) about Section 3. Similarly, you should write some introductory paragraphs for Section 3.1, before you create and use Section 3.1.1. This is exactly the top-down refinement you apply at all levels in the paper.

The writing process itself can be a top-down refinement process. When our graduate students are ready to write their second or third papers by themselves, we usually work with them on the Abstract carefully, together with a 2-3 level outline (including subsection titles and some very brief contents) for the whole paper. This will make the whole paper logical, consistent, and coherent. Then we ask our students to "fill in" details in each section, also in the top-down manner. This will make it easy for your students to write great papers.

6.6   TIPS FOR PAPER WRITING

Above we discussed a number of language-independent misconceptions in paper writing and how to correct them. Here we offer a number of tips for writing good papers, resulting from our own experiences as well as extensive interactions with other authors and reviewers alike.

6.6.1 Use Certain Words to Signal Readers

When you describe your work at a high level (in the top-down refinement process), use certain keywords to signal readers so they know which part is important and which part can be skipped. The following are some examples. We underline those keywords here for illustration.

"We will first describe our method <u>at a high level</u> here. It consists of four <u>major</u> steps. We will describe them <u>briefly</u> below."

"The <u>intuition</u> of our method is that . . ."

"We will describe a proof <u>sketch</u> first to give some <u>intuition</u> about the theorem and the proof."

"<u>Generally speaking</u>, our method is . . . ."

When you start describing technical details, after a high-level description, you should also signal readers about this. You can use keywords such as:

"More specifically, . . ."

"For example, . . ."

"We provide <u>details</u> of Step 3 as follows. . . ."

"Below we give a <u>detailed breakdown</u> of our experimental results."

These words basically signal reviewers that if you are comfortable with the general idea, you can skip the details here!

6.6.2 Use Simple Sentences, But Not Simpler

Einstein once famously said: "your theory should be as simple as possible, but not simpler." The same can be said in writing papers. Often, when beginning researchers write papers, they use long and complex sentences to describe their ideas and work.Most of the time, writing can be simplified to improve readability greatly. Here is a paragraph from a draft paper written by a capable non-native English-speaking Ph.D. student:

"There are two basic functionalities that we expect our system to offer: an effective knowledge organization model to represent the panorama of knowledge in levels and scales, named MKN model; a multi-faceted eBook retrieval system based on MKN model to ease the searching and cognitive process for users, named MIQS, using Facet Grouping, Multi-scale Relevance Analysis, and Information Visualization to overcome the difficulties above."

You will notice that the whole thing is just one sentence! Maybe grammatically it is correct, but it is certainly hard to follow. Interestingly, when we asked the student why he wrote such a long and complex sentence, he replied that he had been preparing for the Graduate Record Examinations (GRE) test. It is a required standardized test for admission to many graduate schools, and it includes tests for reading comprehension. Do you really want to test reviewers' reading comprehension level with your paper?

We revised the long and complex sentence to the following eight short and simple ones:

> "There are two basic functionalities that we expect our system to offer. The first functionality is an effective knowledge organization model. It represents the panorama of knowledge in levels and scales. We call it the MKN model. The second functionality is a multi-faceted eBook retrieval system. It is based on the MKN model to ease the searching and cognitive process for users. We call it MIQS. It uses facet grouping, multi-scale relevance analysis, and information visualization to overcome the difficulties mentioned earlier."

Reviewers will certainly appreciate simple, straightforward, and logical sentences in your paper.

Even when the ideas, methods, or results you want to express are complex, you should try your best to explain them using as simple ways as possible. Often figures and diagrams can be created to help this effort, and analogies can be drawn to simplify the explanation.

6.6.3   Use a Small Set of Terms Throughout the Paper

Another important method to make your paper easy to understand by reviewers and readers is to use a small set of technical terms throughout the paper. Often several technical terms are equivalent in meaning, and different authors may use different terms in their paper. Unlike writing novels and poems in which you would try to avoid using the same terms, in research papers, it is actually better to keep using the same technical terms. Using different terms for the same thing would confuse readers greatly. For example, in the Information Retrieval area, a research field that concerns how documents can be indexed and searched effectively, we often talk about the "precision on the test data." But other terms are also used in other papers, such as "precision on testing data," "precision on validation set," "test-set precision," "percent of correct retrieval on test set." If you use all of these interchangeably in your paper, you will cause a huge amount of confusion.

Thus, it is best to stick to one term for the same meaning and use it throughout the whole paper. If your choice is unconventional, you can formally or informally define the terms early in the paper, or even include a table of term definitions early in the paper, and then use them throughout the paper.

6.6.4   Use Examples Early and Use Them Throughout the Paper

When you explain some abstract concepts or complex modeling processes, it is often best to use one or two concrete examples. Use examples early in the paper. That is, do not overwhelm readers with too many abstract concepts and complex processes before showing examples. By then, the reviewers may have gotten lost (and worse, they may lose interest in your paper). A good approach is to introduce one or two examples early, and to further develop the same example(s) throughout the paper. This is why they are often called "running examples." You can see that we have been using several graduate students as examples throughout the paper, and hopefully our approach has made our book easier and fun to read!

6.6.5   Use Figures, Diagrams, Charts, Photos, Tables

Just as examples can greatly enhance the comprehension of your paper, so too can diagrams, flow charts, photos, figures, drawings, illustrations, tables, and so on. Use them lavishly in the paper. You will find that we have used many figures in this book. "A picture is worth a thousand words." Often it is much faster to view and understand charts and images than reading hundreds of words that explain the same thing. When you make your paper easy to understand and save reviewers' time, they will appreciate it.

6.6.6   Write about Your Motivations and Justifications

As we emphasized earlier, it is extremely important that you write your paper for your reviewers and readers. You must write from their perspective. They may know the area well, but they may know little about the specific research problem studied in your paper. Thus, when you write about your novel theories, methods, processes, and so on, you should write about the motivations—why they work—at least intuitively. If there is a new mathematical equation, explain why you choose its particular form and parameters.

Writing about motivations may not be so easy, as you have worked on your research problem for months or even years. Try to recall your initial motivations, or try to explain your work to some colleagues who do not know your research problem. Or, you can keep track of the motivations and write them down as you go, as part of your overall documentation for your research process, so you can refer to them and look them up later.

Similar to motivations, you must justify certain important choices you made in your research. For example, why did you choose those particular parameters in your experiments and datasets? Why did you combine those processes but not others? For each choice you made, you should try your best to justify it, at least briefly. Reviewers may not "buy" your justifications completely, but if do not see them, they can easily raise questions and cast negative views of your paper.

6.6.7   Pose Potential Questions and Answer Them Yourself

As you are writing for your reviewers who would be reading your paper for the first time, you must think hard about one thing: what questions would they have while reading my paper? These questions can come up anywhere in the paper. For example, in the Introduction, after you briefly describe the problems and your novel solutions, reviewers may wonder why you did not use some previous methods, perhaps with some obvious modifications, that could be effective in solving your problems. In the experiment section, reviewers may wonder why you did not use a different material, method, or a different optimization process.

Writing about motivations and justifications will answer some of those questions. However, sometimes there is no natural place to answer some of these questions. In these cases, *you can pose these questions yourself, at the most likely place that reviewers may ask, and then answer them by yourself*!

The easiest way to pose such questions is "One might wonder . . . ," or "One might argue that . . ." Then provide a succinct answer. If you raise questions at the same time as reviewers would while they are reading your paper, and you provide satisfactory answers, you cannot imagine how satisfied the reviewers will be after reading your paper!

6.6.8   Emphasize and Reiterate Key Points in Your Paper

Often we hear graduate students complaining about reviewers' apparent carelessness when reading their paper. They point out a question from the reviewer but the answer actually appeared in the paper.

As we discussed earlier, reviewers are very busy researchers themselves, and they often have very limited time in reading and reviewing papers. If certain messages are very important, you should emphasize them, and reiterate them several times in the paper, especially in high-level subsections. It may also be worthwhile to present the information in multiple ways, as part of sentences, diagrams, lists, or headings. This allows you to reinforce important information and concepts.How many times and in how many different ways have we reiterated the key point that you should use top-down refinement in writing papers in this book? This should give you an example.

If your method has some small but key differences from the previous methods, and you think reviewers can easily overlook them, you must point this out, and emphasize these key differences. You should also reiterate such key differences several times in the paper, including perhaps in the Introduction and Conclusions.

6.6.9   Make Connections Throughout the Paper

Another important approach, which is often overlooked by young researchers, is to make ample connections within your paper. In our book there are many places we use "See Section . . . for more details," "As we discussed in Section . . . ," "Recall that . . . ," and so on. In fact, whenever you write about something that has been mentioned earlier, or will be mentioned later in the paper, you can consider making a connection there. This makes your paper coherent, consistent, and easy to understand.

6.6.10   Format Papers for Easy Reading

We have presented many approaches and tips on how to make papers easier for reviewers and readers to understand. There are more such tips, but we will not go over them now. Instead, we want to talk about some formatting details. This shows that even "trivial" matters such as formatting should be considered when you write your papers.

One annoying thing we often encounter when we review papers is that figures, charts, and the text explaining them are not on the same page. This can happen easily when you use LaTex™ to format the paper, which arranges the figures or charts automatically. Imagine that reviewers have to flip your paper back and forth to match the text with the diagram—a frustrating process. We often try to reformat the paper manually so the figures and accompanying texts are on the same page, and even near each other.

Another related issue is figure captions and tables. Often you can choose to put explanation texts either in the main text of the paper, or in the captions. We prefer the explanation texts to be put in captions. Captions are always very near (at the top or bottom of) the figures or tables.

Yet another issue is the font size. Often in figures or tables, the fonts are too small to be viewed easily. This is often due to the amount of content presented

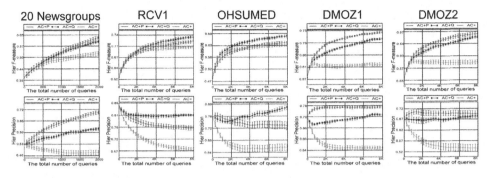

FIGURE 6.2: An example of a paper with too many results on the page.

in the figures and tables. Many authors want to present as much information as possible in figures and tables to show "how much" work they have done. Often too many figures are jammed in a small space, too many curves in each figure, and too many rows and columns in the table. All of these result in small font sizes.

Not only are small fonts hard to view, *with too much information in one figure or table, reviewers can be confused about where to look*! Remember you cannot point out with your finger which column to look at while reviewers are reviewing your paper! Here is an example of a paper with too many results on the page (see in Figure 6.2).

Young researchers are often eager to present as much information as they can fit in the limited space, thinking that perhaps this helps show that they have done a lot of work. In reality, it is often beneficial to only *present a smaller amount of most relevant and key information*, especially in figures and tables, with adequate font sizes.

6.7   OTHER MISCONCEPTIONS AND FLAWS

There are many other misconceptions and flaws in writing research papers. However, as this book focuses mainly on the overall process of doing research for graduate students in science and engineering, we selectively present some key ones here. We plan to write a book exclusively on research paper writing in the future to give this topic complete coverage. Here are some of the other often encountered misconceptions and questions:

- "My work is really elegant and beautiful. It has a lot of math:definitions, theorems, and long proofs. Don't blame me if reviewers cannot understand it!"

- "To emphasize that our method is so much more superior, I use a lot of old, italics, exclamation points, and super strong language in the paper. My supervisor removes most of them. Is he insane or what??!"

- "My supervisor always corrects little things in English that appear everywhere even in newspapers, books (even this book), and magazines. For example, he would change 'it can't . . .' to 'it cannot . . .'. It's really a waste of time, isn't it?"

- "I am putting so many new contributions in the paper, including new theory, proofs, algorithms, experiments, and even an implemented system, so that it cannot be rejected by reviewers."

- "My supervisor asked me to revise the paper further. I think it is already perfect. Besides, I wrote it, so it is hard for me to find my own errors and make further improvement."

- "My work is so much better than previous work, but it still got rejected. I can tell from the reviews that the reviewers are authors, or friends of authors, of the previous work I compared to. How can reviewers accept my paper if their own work is inferior to mine? Is it clearly a conflict of interest?"

- "The reviewers of my journal submission are so unreasonable. I don't think I can ever revise the paper to satisfy their demands. I should try another journal."

6.8   SUMMARY

The two chapters on paper writing (Chapters 5 and 6) are quite long, but they are extremely important for young researchers. Too often young researchers write papers that are unnecessarily hard to understand, and they believe that it is the reviewers' responsibility to understand their papers. Clearly this "flaw" is not related to English or any specific language they are using. A lot of effort can and should be put into writing an easy-to-follow paper that can pass the "10/30" test we mentioned earlier in the chapter. This actually should also be true for writing essays, homework assignments, reports, or any long documents.

Thus, it is not reviewers' responsibility to understand your paper; it is your responsibility to make it easy to understand. You can do that using these approaches and tips we have discussed in Chapters 5 and 6, from the top-down refinement strategy to "trivial" formatting details.

Notice that we have presented two objectives of paper writing: one is to make an argument, as strong as you can, for your paper; the other is to present your arguments and results as clearly and as simply as you can. These two objectives are not conflicting, and a huge amount of time and effort can be put in to optimize them.

Unfortunately, the "global optimal solutions" for these objective functions are unknown, and can never be reached completely. That is, *there are always ways to improve your papers (including research results in the papers)*. The more time and effort you put in, the better the paper will be. Thus, *doing research and paper writing are a never-ending optimization process*. When we submit a paper, it is not because it is already perfect; rather it is because we have put in as much effort as we can before the deadline (most conferences have deadlines), or we believe that the paper is on par with, or better than, the quality of published papers in the same journals or conferences, and we want to put a closure to the current work.

CHAPTER 7

Writing and Defending a Ph.D. Thesis

In previous chapters, we discussed how to conduct effective research and publish research papers. We have discussed how to find good ideas for a thesis in Section 3.5, and how to set up a thesis plan early in Section 3.7. In this chapter, we put all these ideas together and give an overview of one of the most important functions of Ph.D. study: to successfully write and defend a Ph.D. thesis. You will find that writing a thesis is quite similar to writing a research paper (see Chapters 5 and 6), but many important differences also exist.

7.1 THESIS AND DISSERTATION

A Ph.D. thesis or dissertation is the most important piece of work in your early career. It is a logically organized document that convincingly states your solution to a challenging problem. It should be logically organized such that your committee members should understand and appreciate it. The committee of the thesis examiners is tasked to find holes in your thesis, and a Ph.D. defense is organized so that in the space of two or more hours you must bravely and convincingly defend it, in front of the examiners. Your entire graduate life is essentially condensed to this activity!

Sounds tense and terrible? Actually it is not. If we take you as a product, your Ph.D. program as a piece of software, then the difference between when you first entered the program, and when you exit successfully from the program, is a person (you!) who is more confident, logical, knowledgeable in a focused area of research, with knowledge of your problem and solution inside out. When you enter your Ph.D. defense, you are the most knowledgeable person in the proposed area of research (more than most thesis examiners), and you are there to tell your committee just that.

In the traditional sense, a thesis is a hypothesis or statement about the world. For example, a thesis might be that "the Earth goes around the Sun, but not the other way around." Another example of a thesis might be that "a distance measure on the data differences can be found to help build a piece of classification software that performs better than existing classifiers." However, a dissertation is an organized presentation of your thesis, in which you present the thesis and its entire context, present how others have approached the thesis, and your particular solution. A dissertation includes all evidence you have to defend the thesis, and all your thoughts that arise as a result of the thesis. Despite this difference, we often treat them the same. Thus, we often hear advisors and students say, "I am writing my thesis."

In Section 3.5, we briefly discussed the process of going from research ideas to a thesis topic, in which we mentioned that, among the many factors, passion for research, technical strengths, and the area popularity can be important factors for you to consider in choosing a thesis topic. In Section 4.2, we further discussed how to systematically filter the potential thesis ideas based on impact, importance, and knowledge of how to break down a complex problem into simpler ones. When you enter the actual process of writing a Ph.D. thesis, you will find that all of these skills and conditions will play an important role.

The procedures in writing and defending your thesis are all different in different universities and countries, but the main parts are the same. There are basically two ways to organize your thesis: top-down or bottom-up.

7.2   THESIS ORGANIZATION: TOP DOWN OR BOTTOM UP

By the time the student sits down and writes the thesis, typically the student will have accumulated a number of documents as resources. These might include the student's published conference or journal articles, some accepted and some just submitted. They also include the student's proposal documents, and other drafts that the student has written. Now the task for the student is to organize them into a coherent Ph.D. thesis.

First, consider the top-down process (also refer to the top-down process of paper writing in Section 6.4). The top-down approach starts from a central thesis; for example, for a computer science student, a problem such as "How to include a human in the loop when constructing a text classification model?" can be the thesis. It should be simple and concise, and easy to state even to non-domain experts. This is particularly important for you to convince the external Ph.D. committee members when defending your thesis, because by rule the committee must include someone outside of your area of study. For example, a Physics professor might be asked to be the external examiner for a Computer Science Ph.D. student. Then, using the Research Matrix Method that we described in Section 4.3, the thesis might be broken down horizontally or vertically into several subproblems, each with a coherent introduction, existing solution survey, and your proposed solution, along with all experimental and theoretical evidence for you to defend it. Each subproblem can then be further broken down into subsections, each focusing on a sub-subproblem of its own, but all connected in some logical fashion. This is similar to the "thesis plan" we mentioned in Section 3.7.

This top-down approach, illustrated in Figure 7.1, has the following advantages:

- It allows the student and advisor to know about the central thesis early in the program.
- It can easily reveal missing problems and solutions early in the study, so that a remedy can be designed accordingly early in the program.
- It allows the student to have good training in grant proposal writing, and ensures good management style.
- It makes it easy to define coherent terminology and symbols used throughout the thesis.

However, the top-down approach is also not without weaknesses. It is often very challenging for a junior researcher to foresee all subproblems and

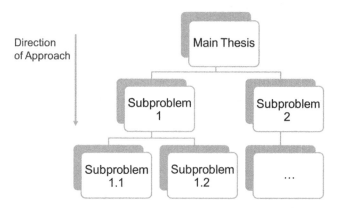

FIGURE 7.1: The top-down approach.

solutions early in the process, as research planning is typically iterative: you plan and then you use the feedback of your plan execution to replan, until the plan is satisfactory.

In contrast to the top-down process of organizing your thesis, a bottom-up process is often favored by many (see Figure 7.2). This applies when the student has published several research papers and has some written resources scattered around. A student using this strategy acts as a gold-miner looking for nuggets in

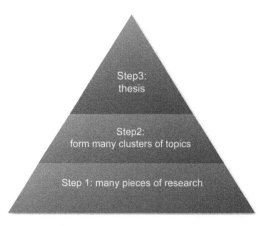

FIGURE 7.2: A bottom-up process for building a thesis.

a pile of rocks. Like a miner, the student has to discover a subset of the material that has a coherent theme in the writing. Again like a miner the student may not take all the raw material in for the content of the thesis, and instead will pick and choose. This is where the student has to have a keen eye on what is necessary to compose the thesis, and what is not. A risk is for the student to mistakenly take the thesis as a job report that includes all that the student has done, thus including some material that has been published by the student, but has little to do with the main theme of the thesis.

Summarizing, the advantages of the bottom-up process are:

- It gives the student more flexibility in exploring different potential themes early in the program.
- The student has a "fat" CV to take to job interviews after completing the thesis, where the diversity adds to advantages in job hunting.

Some of the disadvantages of the approach include:

- There is the risk that the student compiles all published papers as the thesis without selection, thus destroying the coherence of the dissertation.
- The student might spend more time threading together the works reported in papers, and put in much effort unifying the symbols and terminologies used in different papers.

In the end, whether the student prefers a top-down or a bottom-up approach is entirely a personal choice.

Let's consider the earlier examples of Students A2, B2, and C2, where A2 prefers theoretical research, B2 prefers empirical and system-oriented research, while C2 prefers research that can lead to patents and start-up companies. As an example, Student A2 wishes to follow the top-down method for thesis writing and organization, and in so doing he finds a topic to work on first using the matrix method presented in **Chapter 4.** After isolating the thesis topic to be in the area of "transferability for knowledge transfer between domains of different feature spaces," he divides the problem into several subproblems, such as defining a robust distance function, learning a representative subspace, and then transferring between two or more domains with different weights attached to the learning models. These weights are further learned from the source and target data. As a top-down style researcher, Student A2 pursues the problem by breaking it down into finer and finer details.

In contrast, Student B2 pursues his research agenda in a bottom-up manner. Recall that Student B2 is interested in scalable online recommendation algorithms that are both efficient and effective. He publishes several papers on different aspects of the problem, each focusing on a different scalability or accuracy issues associated with his methods. When he accumulates a sufficient number of publications, he then takes stock of what he has accomplished, trying to summarize it into a coherent whole. He finally decides that he will use the material of scalable architecture for online recommendation as his main topic for the thesis, based on which he selects and extends the rest of the subtopics into his thesis as separate chapters.

Finally, Student C2, who favors applications that lead to patents and start-up companies, may select a combination of top-down and bottom-up way of organizing his thesis plan. For example, the student may start by working on a data-mining competition that also leads to several publications and patents. The student then finds the topic of social network analysis to be new and fascinating, and decides to plan for a thesis in this area. He designs a top-down plan to complete the thesis, and works on each part in turn. In the end, the thesis appears like a hierarchically organized "island" of works among several pieces of finished, but different works that are not included in the thesis.

7.3 DEFENDING YOUR THESIS

Before the student celebrates the successful ending of a long student career, there is one more hurdle to cross: the thesis defense. This is where the student has to demonstrate to the experts and non-experts that he or she has proposed a thesis, knows all the previous major works related to the thesis, and has accumulated convincing evidence on the thesis. In addition, the student has to demonstrate that he or she can effectively communicate the thesis.

In many universities, a Ph.D. defense consists of a student presentation (45 minutes to one hour long), followed by a (closed door) question and answer period; the latter part can go in two rounds.

Thesis presentation is unlike any conference or seminar presentations the student has done before, as it summarizes a major project of the student. In a way, your thesis presentation should convince the committee that in the previous five or so years of Ph.D. program, you have contributed to world knowledge by one or more of the following:

- Developed a better solution to an existing challenging problem.
- Identified a new problem and designed its formulation, and offered a new solution, such that in effect you opened up a new area.
- Developed a new methodology that unifies many major approaches in the area, and offers your own new insight.

A thesis presentation should demonstrate that you have understood the problem well including its implications in the real world and that you:

- Know all the previous major approaches to the same problem and can discuss their pluses and minuses with ease.
- Know how to design a solution as well as evaluate its merits and weaknesses.
- Have mastered the skill of presenting the same problem to different audiences, at different lengths and in different contexts.

This last point is particularly important, as presentation skill is one of the distinguishing criteria that separate a Ph.D. from the rest. In this view a Ph.D. must be able to sell his or her thesis, much like a salesperson or a business account manager sells his products. In 40 minutes to 1 hour, the student must convince the committee that the work holds water. The student should be able to summarize the work succinctly in one sentence, three sentences, one paragraph, and several paragraphs, depending on the context in which the presentation is made.

In the actual presentation, the student needs to go through these points:

- what the problem is,
- why the problem is important to solve,
- what the related work solves, how does it solve it, what merits does it have, and what weaknesses does it have.
- how the student has solved it better than others.

These points are similar to the logical steps in any research paper, as described in Section 6.1. When comparing to previous work, you should be as objective as possible, without personal sentiment. Avoid saying overly broad things like, "But their method is ineffective since it cannot scale up."

A thesis presentation is usually followed by an hour-long question-and-answer period. Be prepared, but also be aware that you cannot anticipate all questions. When meeting new and startling questions, pause for a moment before you answer them. Some points to keep in mind:

- Questions can be highly critical from some examiners, but it usually does not mean that they intend to fail you. It is normal that committee members will drill as deeply as they can, in order to see how well you actually know your area. They also wish to see if you can handle tough questions before they let you graduate; after all, they are the "safety guard" for the university's good name.
- Questions from non-experts in a field can be general and be from different, unexpected angles. Don't be afraid to say, "I don't know about this, and I will look into it later."
- Make sure you know the basics of your thesis very well. If examiners find that you misunderstood some basic concepts, they can fail you.
- If you can use slides during a defense, prepare your answers to potential questions with extra slides, just in case questions come up during the Q&A period.

Don't be afraid to pause, think about a question, and discuss how it is relevant to your thesis. You need to be sure you understand what is being asked of you before answering it.

It also helps if you prepare some generic questions that keep popping up in thesis defenses everywhere:

- Summarize your main contributions in one to three sentences.
- What is the problem you are trying to solve?
- What are the main contributions of your thesis? Which one is ground-breaking?
- What is one major weakness of your solution?
- If you were to do your Ph.D. study again, how would you do it differently?

You may still be very nervous before your thesis presentation and defense, especially if your native language is not English. The best advice we can give is to *rehearse it several times before the actual defense.* Ask your supervisor to invite a few colleagues, postdocs, and other graduate students to act as examiners, and stage a "mock defense" for you. As supervisors, we sometimes even video-recorded the whole mock defense process for some of our previous Ph.D. students, and they found it extremely helpful. They became highly confident, and did an excellent job in the thesis defense.

As graduate students, we all hear anecdotes about some famous questions all the time. In the case of one author of this book (Yang), there was a famous professor who was on his committee with a reputation for falling asleep during the student presentation and then waking up, asking: "So, how do you apply this thing?!" This turned out to be the killer question in many of his exams.

. . . . .

CHAPTER 8

Life After Ph.D.

When we were young, becoming a professor seemed to be the ideal thing to do. You become a teacher with your graduate students. You have summer and winter breaks. You get to go to different places for conferences. On top of all this, you are admired by many. In fact, this picture is only partially true. In this chapter, we try to demystify the life of a professor, and give readers a realistic picture of life after Ph.D.

8.1   A DAY IN THE LIFE OF A TYPICAL PROFESSOR

Suppose that you have completed your Ph.D. degree. Now what? One of the authors (Yang) had the following story to tell: after his defense at the University of Maryland, his supervisor Professor Dana Nau congratulated him with a smile: Congratulations! Now you have to write a Ph.D. thesis every year!

This is actually not true. The truth is that after the Ph.D., one typically has to write one Ph.D. thesis every three months! This includes writing grant proposals, preparing and teaching classes, meeting students and mentoring them, serving for conference and journal organizations, attending and sometimes chairing committee meetings, balancing budgets for research groups, and reviewing others' grant proposals. Besides these, you are constantly reading papers, discussing with students, and learning new things. See the list of activities and tasks for researchers in Chapter 1. This seemingly confusing set of roles does not seem so attractive unless they are put in perspective. Here is a picture of a day in the life of a professor in the university.

7:00 a.m.:

> After getting up and having breakfast, and saying goodbyes to family members, the professor heads to the computer to catch up on a few urgent emails that need to be handled right away.

8:00 a.m.:

> The professor heads for the swimming pool or gym. A 30-minute exercise adds infinite energy to the professor's brainpower.

> It is essential for a researcher to have a clear mind when working on a research problem. It is good for anyone to have a healthy body. Students often believe that putting more time on a problem may eventually lead to a desired result, no matter how tired and exhausted they are. However, pouring more time and effort onto a problem is less effective than spending your time and energy wisely. Having at least 30 minutes of physical exercise will allow one to gain much more than spending five times more time on a problem while the mind is murky.

8:30 a.m.:

> The professor heads for office, gets a cup of coffee or tea along the way, and reads more emails and reviews the slides before the 9am class.

9:00 a.m.:

> An undergraduate class starts. A class usually lasts 1.5 hours.

> A faculty member is expected to teach between two to four classes a week. To prepare for the classes, the professor needs to spend more time preparing notes, slides, and other class material, or work with teaching assistants on preparing for assignments, projects, and examinations. Professors are also asked to teach new courses, especially in fast-changing disciplines such as computer science and engineering. In this case, more time needs be spent on the course preparation.

> Teaching is an integral part of a researcher's career. Teaching allows the research to communicate smoothly with students, honing one's ability to explain a difficult concept in plain language, and to illustrate com-

plicated solutions on a variety of easy-to-understand matters. Often professors can also identify exceptional students in the class for collaborative research, and inspire them to become researchers in the future. Communication ability is important for a researcher, as it is part of the researcher's job to write research papers and explain new findings to peers and to people in other fields. It is also important for the researcher to work with more junior researchers and cultivate future generations. Thus, teaching should not be taken as a distraction, but rather, it should be an essential ingredient of being a good researcher.

10:30 a.m.:

During the class, a few students asked questions that required further explanations. The professor invites them to his office to continue the discussion after class.

11:00 a.m.:

The professor joins a committee meeting on graduate admission matters.

A faculty member is expected to do a certain amount of service work. Before getting his or her tenure, a professor is expected to join one or two departmental committees in a light-load mode. After tenure, which typically earns him an associate professor title, the committee work load of a professor is expected to increase, sometimes involving committees beyond his or her own department on matters concerning the school or the university as a whole. The duties of these committees typically involve student admission, scholarship, faculty tenure and promotion, university research grant distribution and proposal ranking, etc. They can range from one to six hours for a senior professor per week.

12:00 p.m.:

The professor has his lunch. Some professors schedule lunch meetings with their students, so that they can discuss in a more relaxed atmosphere, and at the same time, save time.

1:00 p.m.:

The professor attends a student Ph.D. defense session. In his career, a professor is expected to join many students' committees, ranging from undergraduate studies to Ph.D. programs. Sometimes, a professor is called upon to be an external member of a university wide Ph.D. committee, serving on the Ph.D. defenses for students in other departments. On these occasions, he may have to listen to the presentations of members outside their area of expertise.

3:00 p.m.:

The professor meets his Ph.D. students in a group meeting. This is when the professor has the most fun. The professor first listens to a student's presentation of recently conducted experiments, and then raises several questions and discusses possible answers with the students. The professor then works out the next steps in their research with the students, and agrees to send the students some references related to the topic of discussion.

4:30 p.m.:

The professor joins his students and colleagues to attend a departmental seminar. Attending seminars is a typical activity of an academic. This is how professors can learn about others' work, and how they can disseminate their work to others. Sometimes, these seminars are given by job applicants, in which case the professor will be asked to interview and rank the candidates.

5:30 p.m.:

Proposal writing: a professor has to constantly apply for new projects and grants in order to support his group of students, postdoc fellows, and research assistants, and to support his group members when they go attend conferences. All research activities, ranging from computer usage to travel related to conferences, need support. Thus, it is critical that the professor writes successful grant proposals, and writes them often. This is because a typical grant application has a low success rate that can range between 5% to 25%, depending on where you are in the world and the type of grant you apply for. We discuss some tips on how to write a successful grant proposal in Section 8.2.

6:30 p.m.:

The professor returns home for dinner with his family.

8:30 p.m.:

The professor starts working on a research paper that he jointly writes with his students or colleagues. This may be mixed with phone calls, emails, Skype, or in person meetings in research labs.

10:30 p.m.:

The professor starts to book his trip to attend the next conference. He sends emails to his travel agent . . .

As you can see, the professor's day is full of different types of activities, all centered around learning and helping others learn. There are also times when there are too many of these activities that demand attention all at once, which can be distracting. Because of this, a good professor should be highly effective, and also be a good time manager.

Of course, there is more freedom associated with being a researcher. When the class is finished and papers are submitted, most professors can relax a bit. They can also go to different conferences, visit other universities, and chat with other researchers on ideas for future research. This brings us back to Chapter 1, where we discussed the pros and cons of being researchers.

8.2   APPLYING FOR RESEARCH GRANTS

One of the major activities in a professor's career is to apply for research grants. As we described before, these grants are the major funding sources for supporting all activities around a research project, including paying for students' and staff members' salary, lab equipment, computers and printers, and travel expenses to conferences and visits. In the United States, grants are also essential in paying the summer salaries of a professor, since many universities pay for only the teaching terms of a professor's salary. Research grants are typically applied through granting agencies. These granting agencies are typically divided into the following categories:

- University grants: these are the internal grants where a small amount of funds are made available typically to allow new faculty members to get started, or to encourage faculty members to take on a new research direction.

- Government funding agency: these include the National Science Foundation in the US and in China, NSERC in Canada, A-star funding in Singapore, European union grants, and Research Grants Commission in Hong Kong and National Science Foundation of China, and the European Research Council, to name a few.

- Industrial or military research grants: These are sponsored by industry or the military organization, usually with a particular mission in mind. The objectives of these grants or projects are usually more specifically focused on building of prototype products. Stringent constraints on spending and milestones apply.

For a researcher, applying for research grants is a part of everyday life. For most researchers, the size of the grant is an important factor that determines how far you can reach in achieving your research goal. To be a successful grant applicant, the researcher has to be more than a scientist or engineer; he or she must also be a good market researcher in order to understand the needs of the society, a good salesman to present his ideas well, a good writer and presenter, and a good manager in case the project involves more parties. For one person to have all these qualities is challenging, but a successful grant applicant must possess at least some of these attributes.

Often for each type of grant, grant proposals are called for once a year. Applying for a project is similar to submitting a paper to a conference, in that both grants and conferences have deadlines. However, major differences exist. This is underscored by the fact that, often, we hear complaints from even seasoned researchers: "if I have been very successful in getting papers accepted by very prestigious conferences and journals, why do I still fail to get my proposals accepted?!" This is a valid question: why is writing good grant proposals so hard?

The key to understanding the difference between research papers and proposals is that their readership can be very different. While a research paper at a prestigious venue is often reviewed by two to three very qualified experts in one's field of study, a proposal is often reviewed by a much larger pool of people, where only a small subset of them are domain experts and the rest are from a more diverse range of areas (although still in the general area of one's field). Most of the others are there to help assess the general impact and quality of the proposal, and while these reviewers are known in the general areas of research, they may not be experts with the same depth as the author. Thus, if they cannot understand your proposal, the proposal will be ranked low.

Another difference between a proposal and a research paper is that a proposal must motivate the goals and convince the reviewers that the goals are both grand and achievable, without going too much to either extreme. On the one hand, a proposal has to show that the goals to be achieved are challenging, such that few others have tried them before, and thus the goals are innovative. On the other hand, these goals are not so out of reach, at least for this researcher, that the means to reach the goals are viable. This is a fine balance of two competing extremes, and indeed it is very difficult to attain. However, for a research paper, reviewers have the problem and solution in front of him or her from the content of the research paper, thus the balancing act between motivating the problem and solving the problem is less of a problem. In contrast, for a research proposal, it is often impossible to reveal all details of the solution, since otherwise the need for the proposal would be in question.

One of the most important aspects of a research grant proposal is its title. While short, a title is truly a window into the whole proposal. A good title is critical in helping the reviewer form a first impression on the entire proposed project. Recall that in many cases, reviewers themselves are not necessary specific domain experts in the proposed domain; for example, a computer architecture researcher may be called upon to review a proposal on data mining. Thus, the title part should also serve well for the generalist. A rule of thumb is that the title should contain several components, including the target problem (e.g., a learning algorithm for image understanding), the proposed method or solution (e.g., a Bayesian method) and the application area in which the method will be used to solve the problem (e.g., images and text from social media or social networks). A title such as "Image Understanding based on Bayesian Methods for Socially Tagged Images Under Uncertainty and Incomplete Data Environments" might not be a good title, because it is a bit too long. In contrast, "Image Understanding for Socially Tagged Images Under Uncertainty" is more succinct and better.

Similar to the title, the abstract and objectives part of the proposal provides a slightly more detailed window into the proposal, where the aim is to guide the reviewer into the proposal in a more structured manner once the reviewer has been convinced that the proposal is sufficiently interesting by reading the title. Thus, the abstract should state the problem, tell the reader why these problems are challenging and new, and then state in general terms what the proposed solutions are. This is followed by a few sentences on the impact of the solution should the project be funded and succeed. We have given similar advice on paper writing in Chapters 5 and 6.

In a similar vein, the objective part of a proposal should state clearly what the proposed tasks and aims are for the project to accomplish. If there are several objectives in the statement, it is important that there is a main objective, which, like the title, should be designed so as to impress the reader. This main objective can be stated at the beginning of the list of objectives, or at the end. It may often be the case when the writer of the proposal has done some preliminary works, that he might wish to use the first objective as a lead-in to the rest of the proposal. This is fine as long as the researcher has an overview sentence to tell the reader about the case. If the main objective is stated first, then the rest are sub-objectives and should be grouped as subtasks under the first important objective. This forms a top-down taxonomy for the reader to follow.

In general, a research proposal is broken down into several functional sections, each serving a different purpose:

- The abstract explains the proposed project even to an outsider, making clear in one or two sentences the main problem, the target audience, the open challenges and open problems, motivations related to these challenges and problems, proposed solutions at a high level, and the impact your solution will bring should the project be funded and succeed.

- An overview of the objectives that describes the main things you wish to achieve, along with an indication that your work is important not only for your own research field, but for society as a whole. The objectives should be hierarchically organized so that readers can follow each path to other important parts of the rest of the proposal. Readers can use this to conduct a review of the proposal in a non-sequential manner, jumping to parts that they wish to see efficiently.

- A review of previous works and background research that describes the context of the proposed project as well as what the investigators and other people have done in the proposed research area in the past. It is important to stay focused in the background discussion so as not to diverge into too many irrelevant details and marginally related fields. The main purpose of this part is, first, to convince the readers that the researchers are true experts in the field of study; second, to show that there are not many prior works addressing the same problem being proposed; and third, to demonstrate that the investigators themselves have done preliminary work leading to the proposed projects.

- The methodology and research plan part, discusses in detail the research method, subtasks, research schedule, and evaluation methods.

Important in this part are the references back to previous sections that discuss specific research subtasks and steps to the keywords indicating the major objectives in the title, abstract, and objectives part; these can be served by sentences such as "recall in Section 1, we discussed the importance of designing a new method for ABC; in this section we describe the details of ABC." The more tightly connected these parts are via references like this, the easier it is for reviewers to master the main points of the proposal, and thus the better the outcome.

- The budget and research schedule part, should be carefully proposed rather than going to either extremes of asking for too much funding or too little. In one extreme, if too much funding is requested, it is difficult to justify the proposal in terms of its scope. But if too little is asked, reviewers will question the feasibility of the proposed project as well.

To help the reader understand the full scope of reviewing criteria, here we quote the proposal review criteria from the U.S. National Institute of Health (NIH), a well-known funding agency, as an example. This set of criteria is fairly typical of all criteria of many major grant agencies across the world[1]:

- **Overall Impact.** Reviewers will provide an overall impact/priority score to reflect their assessment of the likelihood for the project to exert a sustained, powerful influence on the research field(s) involved, in consideration of the following review criteria and additional review criteria (as applicable for the project proposed).
- **Significance.** Does the project address an important problem or a critical barrier to progress in the field? If the aims of the project are achieved, how will scientific knowledge, technical capability, and/or clinical practice be improved? How will successful completion of the

---

[1] See http://grants.nih.gov/grants/peer/critiques/sbir-sttr.htm#sbir-sttr_01

aims change the concepts, methods, technologies, treatments, services, or preventative interventions that drive this field? Does the proposed project have commercial potential to lead to a marketable product, process, or service?

- **Investigator(s).** Are the PD/PIs (Project Directors and Principle Investigators), collaborators, and other researchers well suited to the project? If Early Stage Investigators or New Investigators, or in the early stages of independent careers, do they have appropriate experience and training? If established, have they demonstrated an ongoing record of accomplishments that have advanced their field(s)? If the project is collaborative or multi-PD/PI, do the investigators have complementary and integrated expertise? Are their leadership approach, governance, and organizational structure appropriate for the project?

- **Innovation.** Does the application challenge and seek to shift current research or clinical practice paradigms by utilizing novel theoretical concepts, approaches or methodologies, instrumentation, or interventions? Are the concepts, approaches or methodologies, instrumentation, or interventions novel to one field of research or novel in a broad sense? Is a refinement, improvement, or new application of theoretical concepts, approaches or methodologies, instrumentation, or interventions proposed?

- **Approach.** Are the overall strategy, methodology, and analyses well-reasoned and appropriate for accomplishing the specific aims of the project? Are potential problems, alternative strategies, and benchmarks for success presented? If the project is in the early stages of development, will the strategy establish feasibility and will particularly risky aspects be managed?

- **Environment.** Will the scientific environment in which the work will be done contribute to the probability of success? Are the institutional support, equipment, and other physical resources available to the investigators adequate for the project proposed? Will the project benefit from unique features of the scientific environment, subject populations, or collaborative arrangement?

What are some of the typical mistakes made by the writers of grant proposals? The following is an excerpt of some of the comments summarized from reviewers' comments by Hong Kong's natural sciences and engineering research agency, or the RGC (research grants council).[2] Note that these are examples of generalized (negative) comments from many proposals accumulated in several years:

Originality

- There is only limited innovation.
- This is an incremental research. There is lack of evidence of new breakthrough.
- Objective 1 is not exciting. Objective 2 seems to be a part of Objective 3. Overall the proposal lacks new ideas to investigate.
- The objectives, even if achieved, are not exciting.

Methods

- It is unclear how the data will be collected and analyzed in task 1.
- Difficult to see how the research design/methodology will achieve the project objectives.
- The PI (principle investigator) needs to write the design and methodologies more clearly. For this purpose, step-by-step procedures, diagrams, and equations would be helpful.

Feasibility

- The Principle Investigator (PI)'s track record is a concern to this reviewer, as he has not published any paper or patent in the area of research in this proposal.

[2] http://www.ugc.edu.hk/eng/rgc/index.htm

- This proposal on <topic> was submitted before and was not supported. Some arguments have been provided, but do not seem convincing.
- The PI has limited understanding of <research area>, which is very important to the second objective. The PI should have involved co-Is to contribute missing expertise.
- No preliminary results are provided to support the hypothesis.
- This work is not properly based on existing published knowledge in the subject area.
- The objectives of this proposal are too wide. I believe too much has been promised in this proposal for two years of work.

Likely contribution to discipline

- The PI failed to clarify how the proposed technique can achieve any better performance than other state-of-the-art techniques.
- It is stated that an understanding of <topic> is of paramount importance. You need to describe for whom this is important and why.
- I am not optimistic that the proposed research will generate significant publications in main-stream journals.

Others

- This proposal is full of ambiguities and I do not think the applicant has a clear mind where he is heading. Basic concepts and many technical details are unclear.
- The proposal has three objectives. These objectives are not integrated; they appear disjointed.
- The manpower planning is poorly put together.
- The proposal is too qualitative when it should be a quantitative topic.
- There is no sufficient information in the proposal about the analytical framework, theoretical foundation, model specification, estimation strategy, data source, and potential economic impact to Hong Kong and beyond.

- I don't think the current title properly reflects what the authors intend to do.

8.3   TECHNOLOGY TRANSFER

A professor's life is not just teaching and research. Often times, a professor, and his students and colleagues, might invent something useful for society, and this is when they start thinking about how to move the technology and ideas from their labs to the real world in areas such as industry and the market place. This process of moving knowledge away from labs, conferences, and journals, to practice is known as technology transfer.

A researcher can conduct technology transfer at any point in time. He or she may transfer knowledge while still being a graduate student. He or she may be a postdoc fellow or a professor, or an industrial researcher working in an industry lab such as Google, Microsoft Research, General Electric, Huawei, or pharmaceutical companies such as Pfizer, etc. Technology transfer may take one's time away from pure academic research for a while, but it is a very satisfying process, especially after seeing one's ideas and work being used by people.

There can be several ways in which technology transfer can happen. The simplest way is licensing, where, with the help of a legal consultant, one can sign an agreement with a company such that the company can use the technology in a limited manner. If the technology is a piece of software, then the agreement will state clearly what the software includes and does not include, what happens when the software refers to some other people's software that they themselves have limited use clauses, and what happens when the software breaks. That is why a legal expert is often needed to assist in negotiating and drafting a licensing agreement. The process of defining these terms can be lengthy and tedious. Fortunately, many universities and research labs have offices that are designated for this use, and in many cases this costs the researcher very little or nothing, since the legal services can often recuperate their costs through a percentage of the agreed fee as part of the agreement.

When reading a legal agreement, such as a licensing agreement, for the first time, a researcher is often inundated with jargon and clauses that are quite foreign to the researcher. When this happens, do not worry, since a legal document is not very different from a research paper. Similar to a research article, the documents often start with definitions in upper case; these are the terms that will be used repeatedly and unambiguously in the subsequent texts. The legal terms are similar to a set of logical rules and statements. While a lawyer is helpful, it is also the researchers' task to ensure that these statements are consistent and the coverage is appropriate. Mathematical logic does help here.

An important aspect of licensing is the terms of use of the product to be licensed. The term can be limited by geography, such as the Asian or European market, or by time, such as number of years. In some extreme cases, the researcher might agree to give all rights of use to the company, in which cases the term might refer to "exclusive use," which forbids the researcher to transfer the same technology to others, essentially preventing competitors from having access to the same technology. In such cases, the benefit to the researcher is usually higher than non-exclusive use clauses.

Patents are another kind of technology transfer. Patents are statements of a new invention, be it a process or a product design that is associated with a right assigned to the inventor. When filing patents on an idea, a researcher goes through a process that is almost the same as doing research. First, the researcher should define what the idea is in the simplest and most unambiguous terms. Then, the researcher goes through a literature search to prove that other similar ideas or technology are in fact different from the one concerned, in one way or the other. This is like writing a related work section of an article or a Ph.D. dissertation. Often there are patent databases in a library to help one go through this search process, sometimes with a fee, but many search engines today provide look-up services to the public for free. Like a research article, a patent application should reference many research papers in citations. In addition, the researcher should give the full details on the design of the idea, such as how it might be used in practice, and so on. A patent lawyer, who will assist in the filing process, often reviews this document. The filing time may in fact be quite long, sometimes two years, and the fee can range from thousands of US dollars to tens of thousands. When filing a patent on an idea, there is a requirement that no prior publication has been made on the same idea, thus the researcher is often forbidden to publish a research article on the same idea until the patent is approved by the patent office. In some cases, however, a patent can be filed while a paper is being reviewed by a conference or a journal, but researchers should consult a lawyer about this.

Sometimes a researcher spends some time working for a company on a limited time basis. In such cases, we say that the researcher is consulting with the company. Often, universities encourage researchers to do some consulting work while being employed at a university. For example, many universities in the US, Canada, and Hong Kong allow a faculty member to spend one day per week to consult with a company. Consulting activities vary greatly from person to person. In one case, a faculty might answer questions from the people working in the company that they consult with. In other cases, a faculty member might actually sit in at the company as if he or she was an employee of the company.

The most complicated, but also most rewarding offshoot of technology transfer is spinning off a company. We hear many legends of spin-off companies, such as Google, which was formed by two Ph.D. students who invented a new search technology. Universities often turn on a green light for faculty members and students to go create spin-off companies, by allowing them a "leave of absence." This is a term that refers to the practice of allowing a faculty member to leave the university for a specified time period, often one or two years, in which the researcher is not paid a salary by the university, but the position is kept for the person. This is why the researcher needs to find the funding required to support him or herself during the spin-off company creation process.

Spin-off companies can be created with or without a faculty member. In the case of Student C, for example, the student may decide to form a spin-off company with the help of the university Technology Transfer office, taking advise from the supervisors as well as industrial partners. For example, student C might find his new algorithms to perform much better than previously known algorithms in an e-commerce area, and may decide to create a spin-off company in order to commercialize this algorithm. He and his supervisor may decide to leave the university for a while, and formulate a business plan and marketing plan to extend his algorithms to more industrial applications. His previously filed patent on the algorithm would have helped a lot when talking to potential venture capital companies, many of whom decide to take a portion of the company's share and provide funding. Furthermore, the venture capital companies may have access to a large network of experienced business people, from whom Student C selected a few as the new companies' business development officers, such as Chief Executive Officers (CEOs) or Chief Financial Officers (CFOs). The student might take the position of a Chief Technology Officer (CTO), which allows the student to further expand his talent in product development.

However, what we often don't hear is the amount of work that these successful, or unsuccessful, companies require. While it sounds nice to be able to hand others a business card with a title of CEO or CTO, in fact, before deciding on spinning off a company, the researcher should take many factors into careful consideration, as the risk is also the greatest among all options of technology transfer.

A necessary first step in finding venture capital to support a spin-off company is to write a business plan. When writing a business plan, first and foremost, researchers should describe their objectives in sufficient depth so that funding parties can be sufficiently convinced that there is indeed an untapped market out there for the technology. This process is very much similar to describing one's ideas in a research article. Writing this up is also akin to writing the Introduction section of a paper (see Chapter 6). The researcher himself or herself should be firstly convinced of the existence of the market.

Then, the researcher should prepare a thorough literature study, like building a related work section of their paper, in showing that despite the existence of such a lucrative market, in fact few competitors have tried exactly the same ideas before. This requires a detailed argument and analysis, often with many citations and quotes.

Subsequently, the researcher has to discuss the methodology itself and a business model in which he or she will specify how this company can survive based on the revenue received by selling this product after taking cost into consideration. There is a wide spectrum of operating methods which a company can rely on, ranging from building up a service and sales channel for the product to reach a larger audience, to facilitating a strategic alliance with other companies to mutually benefit from the sales. Note that in a business plan, the technology description part is needed and this part is similar to writing a research paper; however, unlike a research paper, the technology part of a business plan only occupies a small portion, as a strong argument is needed in analyzing the market and the competitors. It is sometimes estimated that technology often occupies only five to ten percent of an entire business building process. Even so, the reward far out-runs the hurdles, since otherwise we would not have seen so many successful examples.

There is also the management aspect to consider. At this point, the researcher has to think like a manager, as a team of experts will be needed to work in sync. This team is known as a "management team," which includes not only the researcher, but a person with business management experience as a CEO, a financial and accounting expert known as a CFO, a publicity expert known as a Chief Information Officer (CIO), etc. The team is in fact the most critical part of the spin-off company, and because of this, a venture capital company sometimes will assemble such a team for the researcher, by inviting other experienced people to join in. This is when the researcher will see his or her own shares in the company shrink, but in fact the whole pie might become much larger. Thus it is a worthwhile practice in many cases. With the help of these experts, the researcher will further complete a financial plan and a marketing plan for the spin-off company.

One of the stickiest issues when creating a spin-off company is the IP, or intellectual property. This refers to the content of the technology, a specification of who owns it and a claim that it has never been done before. In some universities and research labs, the university owns partial IP for any invention created therein. When creating a company, legally the university can claim a portion of the company's share. In this case, it is important for the researcher to negotiate with the company early in the process, so that the university or the research lab also bears a part of the cost in early stages of the company's creation, such as providing subsidized office space, computing facilities, and other legal services.

Summary

In this book, we have systematically discussed how to do research, from setting one's goals for a research career, to getting research ideas, reading and critiquing papers, formulating a research plan, writing and publishing research papers, and to writing and presenting one's thesis. We also discussed what life is like after one gets a Ph.D. degree. We hope the examples, lessons, and experiences presented in this book demystify a researcher's life, so that young and inspiring students can set the right goals in life, and young and beginning researchers have something to rely on. We will succeed if this book can offer some guidance to students and junior researchers alike to help them succeed in a fruitful and adventurous research life.

Indeed, research is full of adventures and fun if you master the secrets of doing it right. In our life we totally enjoyed doing research ourselves. We certainly hope that you do too.

. . . .

References

[1] Advice on Research and Writing. Mark Leone. http://www.cs.cmu.edu/~mleone/how-to.html

[2] Advice to a Beginning Graduate Student. Manuel Blum. http://www-2.cs.cmu.edu/%7Emblum/research/pdf/grad.html

[3] "A Ph.D. is Not Enough." Peter J. Feibelman. Addison-Wesley, Reading, MA, 1993.

[4] The Craft of Research (Guides to Writing, Editing, and Publishing). Wayne C. Booth, Gregory G. Colomb, and Joseph M. Williams, The University of Chicago Press.

[5] How to Research. Loraine Blaxter, Christina Hughes, and Malcolm Tight. Open University Press.

[6] 'How to Write a Research Paper.' Martyn Shuttleworth, LuLu.com. August 2010.

[7] Surely You're Joking, Mr. Feynman!: Adventures of a Curious Character, Richard Feynman, Ralph Leighton (contributor), and Edward Hutchings (editor). W W Norton, 1985.

[8] Advice to a Young Scientist. Peter Brian Medawar. Harper & Row, 1979.

Author Biographies

Charles Ling has been a professor at the Western University in Canada since 1989. He obtained his BSc. in Computer Science from the Shanghai Jiaotong University in 1985, and then graduated with his Masters and then Ph.D. in Computer Science from the University of Pennsylvania, USA, in 1987 and 1989, respectively. He specializes in data mining and machine learning, and their applications in Internet, business, healthcare, and bioinformatics. Overall, he has published over 120 research papers in peer-reviewed conferences and journals. Charles Ling has also engaged in much professional service in the above areas. He has been an associate editor for several top journals and an organizer for several top conferences in computer science. He is also a Senior Member of IEEE and a Lifetime Member of AAAI (Association of Advancement of Artificial Intelligence). Charles Ling is the director of the Data Mining and Business Intelligence Lab at Western University, where has been involved in several technology transfer projects.

Charles Ling is also a specialist in child gifted education. He integrates his research in Artificial Intelligence and cognitive science, and develops a full range of thinking strategies that improve children's intellectual abilities. These thinking strategies embrace, enhance, and connect with math, science, and other areas. He can be reached at cling@csd.uwo.ca.

Qiang Yang is a professor in the Department of Computer Science and Engineering, Hong Kong University of Science and Technology, Hong Kong, China. He is an IEEE Fellow, IAPR Fellow (International Association of Pattern

Recognition) and an ACM Distinguished Scientist, for his contributions to Artificial Intelligence and Data Mining, which are also his research interests.

Qiang Yang graduated from Peking University in 1982 with BSc. degree in Astrophysics, and obtained his PhD degree in Computer Science from the University of Maryland, College Park, in 1989. He was an assistant and then associate professor at the University of Waterloo, Canada between 1989 and 1995, and an associate and then a full professor and NSERC Industrial Research Chair at Simon Fraser University in Canada from 1995 to 2001. He is an author of two books and over 300 publications on AI and data mining. His research teams won several prestigious international competitions on data mining. He was an invited speaker at several top conferences, and a founding Editor in Chief of an ACM journal (ACM TIST), as well as an editor for several top journals. He has also been an organizer for some top conferences in computer science. Besides academic research, he has engaged in several industrial projects with IT companies, and been sitting on several research grant panels. In his spare time, he enjoys sports, reading and traveling. He can be reached at qyang@cse.ust.hk.

Made in United States
North Haven, CT
09 February 2022

15928071R00096